HODDER GIBSON

IN ASSOCIATION WITH

SQA

Model Papers

WITH ANSWERS

PLUS: Official SQA Specimen Paper
With Answers

Advanced Higher for CfE
Chemistry

2015 Specimen Question Paper
& Model Papers

HODDER GIBSON
AN HACHETTE UK COMPANY

Hodder Gibson is grateful to the copyright holders, as credited on the final page of the Answer Section, for permission to use their material. Every effort has been made to trace the copyright holders and to obtain their permission for the use of copyright material. Hodder Gibson will be happy to receive information allowing us to rectify any error or omission in future editions.

Hachette UK's policy is to use papers that are natural, renewable and recyclable products and made from wood grown in sustainable forests. The logging and manufacturing processes are expected to conform to the environmental regulations of the country of origin.

Orders: please contact Bookpoint Ltd, 130 Park Drive, Milton Park, Abingdon, Oxon OX14 4SE. Telephone: (44) 01235 827720. Fax: (44) 01235 400454. Lines are open 9.00–5.00, Monday to Saturday, with a 24-hour message answering service. Visit our website at www.hoddereducation.co.uk. Hodder Gibson can be contacted direct on: Tel: 0141 848 1609; Fax: 0141 889 6315; email: hoddergibson@hodder.co.uk

This collection first published in 2015 by
Hodder Gibson, an imprint of Hodder Education,
An Hachette UK Company
2a Christie Street
Paisley PA1 1NB

Typeset by Aptara, Inc.

Printed in the UK

A catalogue record for this title is available from the British Library

ISBN: 978-1-4718-6047-8

3 2 1

2016 2015

Introduction

Study Skills – what you need to know to pass exams!

Pause for thought

Many students might skip quickly through a page like this. After all, we all know how to revise. Do you really though?

Think about this:

"IF YOU ALWAYS DO WHAT YOU ALWAYS DO, YOU WILL ALWAYS GET WHAT YOU HAVE ALWAYS GOT."

Do you like the grades you get? Do you want to do better? If you get full marks in your assessment, then that's great! Change nothing! This section is just to help you get that little bit better than you already are.

There are two main parts to the advice on offer here. The first part highlights fairly obvious things but which are also very important. The second part makes suggestions about revision that you might not have thought about but which WILL help you.

Part 1

DOH! It's so obvious but ...

Start revising in good time

Don't leave it until the last minute – this will make you panic.

Make a revision timetable that sets out work time AND play time.

Sleep and eat!

Obvious really, and very helpful. Avoid arguments or stressful things too – even games that wind you up. You need to be fit, awake and focused!

Know your place!

Make sure you know exactly **WHEN and WHERE** your exams are.

Know your enemy!

Make sure you know what to expect in the exam.

How is the paper structured?

How much time is there for each question?

What types of question are involved?

Which topics seem to come up time and time again?

Which topics are your strongest and which are your weakest?

Are all topics compulsory or are there choices?

Learn by DOING!

There is no substitute for past papers and practice papers – they are simply essential! Tackling this collection of papers and answers is exactly the right thing to be doing as your exams approach.

Part 2

People learn in different ways. Some like low light, some bright. Some like early morning, some like evening / night. Some prefer warm, some prefer cold. But everyone uses their BRAIN and the brain works when it is active. Passive learning – sitting gazing at notes – is the most INEFFICIENT way to learn anything. Below you will find tips and ideas for making your revision more effective and maybe even more enjoyable. What follows gets your brain active, and active learning works!

Activity 1 – Stop and review

Step 1

When you have done no more than 5 minutes of revision reading STOP!

Step 2

Write a heading in your own words which sums up the topic you have been revising.

Step 3

Write a summary of what you have revised in no more than two sentences. Don't fool yourself by saying, "I know it, but I cannot put it into words". That just means you don't know it well enough. If you cannot write your summary, revise that section again, knowing that you must write a summary at the end of it. Many of you will have notebooks full of blue/black ink writing. Many of the pages will not be especially attractive or memorable so try to liven them up a bit with colour as you are reviewing and rewriting. **This is a great memory aid, and memory is the most important thing.**

Activity 2 – Use technology!

Why should everything be written down? Have you thought about "mental" maps, diagrams, cartoons and colour to help you learn? And rather than write down notes, why not record your revision material?

What about having a text message revision session with friends? Keep in touch with them to find out how and what they are revising and share ideas and questions.

Why not make a video diary where you tell the camera what you are doing, what you think you have learned and what you still have to do? No one has to see or hear it, but the process of having to organise your thoughts in a formal way to explain something is a very important learning practice.

Be sure to make use of electronic files. You could begin to summarise your class notes. Your typing might be slow, but it will get faster and the typed notes will be easier to read than the scribbles in your class notes. Try to add different fonts and colours to make your work stand out. You can easily Google relevant pictures, cartoons and diagrams which you can copy and paste to make your work more attractive and **MEMORABLE**.

Activity 3 – This is it. Do this and you will know lots!

Step 1

In this task you must be very honest with yourself! Find the SQA syllabus for your subject (www.sqa.org.uk). Look at how it is broken down into main topics called MANDATORY knowledge. That means stuff you MUST know.

Step 2

BEFORE you do ANY revision on this topic, write a list of everything that you already know about the subject. It might be quite a long list but you only need to write it once. It shows you all the information that is already in your long-term memory so you know what parts you do not need to revise!

Step 3

Pick a chapter or section from your book or revision notes. Choose a fairly large section or a whole chapter to get the most out of this activity.

With a buddy, use Skype, Facetime, Twitter or any other communication you have, to play the game "If this is the answer, what is the question?". For example, if you are revising Geography and the answer you provide is "meander", your buddy would have to make up a question like "What is the word that describes a feature of a river where it flows slowly and bends often from side to side?".

Make up 10 "answers" based on the content of the chapter or section you are using. Give this to your buddy to solve while you solve theirs.

Step 4

Construct a wordsearch of at least 10 × 10 squares. You can make it as big as you like but keep it realistic. Work together with a group of friends. Many apps allow you to make wordsearch puzzles online. The words and phrases can go in any direction and phrases can be split. Your puzzle must only contain facts linked to the topic you are revising. Your task is to find 10 bits of information to hide in your puzzle, but you must not repeat information that you used in Step 3. DO NOT show where the words are. Fill up empty squares with random letters. Remember to keep a note of where your answers are hidden but do not show your friends. When you have a complete puzzle, exchange it with a friend to solve each other's puzzle.

Step 5

Now make up 10 questions (not "answers" this time) based on the same chapter used in the previous two tasks. Again, you must find NEW information that you have not yet used. Now it's getting hard to find that new information! Again, give your questions to a friend to answer.

Step 6

As you have been doing the puzzles, your brain has been actively searching for new information. Now write a NEW LIST that contains only the new information you have discovered when doing the puzzles. Your new list is the one to look at repeatedly for short bursts over the next few days. Try to remember more and more of it without looking at it. After a few days, you should be able to add words from your second list to your first list as you increase the information in your long-term memory.

FINALLY! Be inspired...

Make a list of different revision ideas and beside each one write **THINGS I HAVE** tried, **THINGS I WILL** try and **THINGS I MIGHT** try. Don't be scared of trying something new.

And remember – "FAIL TO PREPARE AND PREPARE TO FAIL!"

Advanced Higher Chemistry

The course

The main aims of the Advanced Higher Chemistry course are for learners to:

- develop a critical understanding of the role of chemistry in scientific issues and relevant applications, including the impact these could make on the environment/society

- extend and apply knowledge, understanding and skills of chemistry

- develop and apply the skills to carry out complex practical scientific activities, including the use of risk assessments, technology, equipment and materials

- develop and apply scientific inquiry and investigative skills, including planning and experimental design

- develop and apply analytical thinking skills, including critical evaluation of experimental procedures in a chemistry context

- extend and apply problem-solving skills in a chemistry context

- further develop an understanding of scientific literacy, using a wide range of resources, in order to communicate complex ideas and issues and to make scientifically informed choices

- extend and apply skills of independent/ autonomous working in chemistry.

How the course is assessed

To gain the course award:

- You must pass three units which are assessed internally on a pass/fail basis. The units are:
  1. Inorganic and Physical Chemistry
  2. Organic Chemistry and Instrumental Analysis
  3. Researching Chemistry

- You must submit a project which is marked externally by the SQA and is worth a total of 30 marks.

- You must sit the Advanced Higher Chemistry exam paper which is set and marked by SQA and is worth 100 marks.

The course award is graded A–D, the grade being determined by the total mark you score in the examination and the mark you gain in the project.

The examination

The examination consists of one paper which has two sections and lasts 2 hours 30 minutes. Section 1 is multiple choice and is worth 30 marks. Section 2 comprises extended answer questions and is worth 70 marks.

Further details can be found in the Advanced Higher Chemistry section on the SQA website: http://www.sqa.org.uk/sqa/48459.html

Key tips for your success

Practise! Practise! Practise!

The key to exam success in Chemistry is to prepare by answering questions regularly. Use the questions as a prompt for further study. If you find that you cannot answer a question, review your notes and/or textbook to help you find the necessary knowledge to answer the question. This is a very quick way of finding out what you can or cannot do and is a better use of your time than copying notes passively, which is a common trap many students fall into!

The data booklet

The data booklet contains formulae and useful data which you will have to use in the exam. Although you might think that you have a good memory for chemical data (such as the symbols or atomic masses of elements), it is always worthwhile to check using the data booklet.

Calculations

In preparation for the exam, ensure that you recognise the different calculation types and know when to use the formulae from the data booklet:

- Energy, frequency and wavelength

$$E = Lhf$$

$$c = f\lambda$$

- Using the pH equations

  strong acids: $pH = -\log[H^+]$

  weak acids: $pH = \frac{1}{2}pK_a - \frac{1}{2}\log c$

  buffers: $pH = pK_a - \log\frac{[acid]}{[salt]}$

- Thermodynamics

$$\Delta S = \Sigma S^0 \text{products} - \Sigma S^0 \text{reactants}$$

$$\Delta H = \Sigma H^0 \text{products} - \Sigma H^0 \text{reactants}$$

$$\Delta G^0 = \Delta H^0 - T\Delta S^0$$

Remember that you have to make the units for ΔS compatible with ΔH and ΔG. This can be done by dividing ΔS by 1000 so that it has the units $kJ K^{-1} mol^{-1}$ **or** by multiplying both ΔH and ΔG by 1000 so that they have the units $J mol^{-1}$.

- Kinetics: calculating the rate constant, k, from rate equations
- Synthesis: calculating the % yield
- Solutions: calculating % solution by mass, % solution by volume and ppm
- Calculations from balanced chemical equations (stoichiometry)
- Calculating the empirical formula

You will encounter these calculations in the exam so it is worth spending time practising them as part of your revision to ensure that you are familiar with the routines for solving these problems. Even if you are not sure how to attempt a calculation question, show your working! You will be given credit for calculations which are relevant to the problem being solved.

"Explain" questions

You will encounter questions which ask you to *explain your answer*. Take your time when answering such questions and try to give a clear explanation. If possible, use diagrams or chemical equations; they can really bring an answer to life.

Applying your knowledge of practical chemistry

As part of your Advanced Higher Chemistry experience, you should have had plenty of practice carrying out experiments using standard lab equipment. You will also have had opportunities to evaluate your results from experiments. In the Advanced Higher exam, you are expected to be familiar with the techniques and apparatus given in the following table. This table has been taken from the SQA AH Chemistry Course Support Notes, April 2015 (http://www.sqa.org.uk/files_ccc/AHCUSNChemistry.pdf).

Learners would benefit from being familiar with the following apparatus, practical techniques and activities:	Learners should be able to process experimental results by:
<ul><li>digital balance</li><li>Buchner or Hirsch or sintered glass funnel</li><li>glassware with ground glass joints ('Quickfit' or similar)</li><li>thin-layer chromatography apparatus</li><li>weighing by difference and gravimetric analysis</li><li>preparing a standard solution</li><li>using a reference or control or blank determination</li><li>carrying out a complexometric titration</li><li>carrying out a back titration</li><li>using a colorimeter or visible spectrophotometer and carrying out dilution to prepare a calibration graph</li><li>distilling</li><li>refluxing</li><li>using vacuum filtration methods</li><li>recrystallising</li><li>determining % yield experimentally</li><li>using thin-layer chromatography</li><li>using melting point apparatus and mixed melting point determination</li><li>using a separating funnel and solvent extraction</li></ul>	<ul><li>representing experimental data using a scatter graph</li><li>sketching lines or curves of best fit</li><li>calculating mean values for experiments</li><li>identifying and eliminating rogue points from the analysis of results</li><li>qualitative appreciation of the relative accuracy of apparatus used to measure the volume of liquids (Learners would be expected to know that the volume markings on beakers provide only a rough indication of volume. While measuring cylinders generally provide sufficient accuracy for preparative work, for analytic work, burettes, pipettes and volumetric flasks are more appropriate.)</li><li>appreciating that when a measurement has been repeated, any variations in the value obtained give an indication of the reproducibility of the technique</li><li>knowing that the uncertainty associated with a measurement can be indicated in the form, *measurement ± uncertainty* (Learners are not expected to conduct any form of quantitative error analysis.)</li><li>quantitative stoichiometric calculations</li><li>interpretation of spectral data</li></ul>

Analysis of data

From your experience of working with experimental data in both Higher and Advanced Higher Chemistry, you should know how to calculate averages, eliminate rogue data and how to draw and interpret graphs (scatter and best fit line/curve).

It is common in Chemistry exams to be presented with titration data such as shown in the following table.

Titration	Volume of solution/cm^3
1	26·0
2	24·1
3	39·0
4	24·2
5	24·8

Result 1 is a rough titration which is not accurate.

Results 2 and 4 could be used to calculate an average volume (= 24·15 cm^3).

Result 3 is a rogue result and should be ignored.

Result 5 cannot be used to calculate the average volume as it is too far from 24·1 and 24·2. In other words, it is not accurate.

Open-ended questions

Real-life chemistry problems rarely have a fixed answer. In the Advanced Higher exam you will encounter two three-mark questions that are open-ended; this means that there is more than one "correct" answer. You will recognise these from the phrase *using your knowledge of chemistry* in the question. *To tackle these questions, focus on the chemistry.* For example, if you are shown a molecule, look out for familiar functional groups. Perhaps you could answer the question by discussing the typical reactions and properties of these functional groups. If it is an organic molecule, you could discuss spectroscopic techniques and the likely results you would get if you were to analyse the molecule by NMR or IR. Other questions will require more specific knowledge of the AH course. Treat these as an opportunity to show the examiner how much you know about the topic of interest. However, ensure your answer actually answers the question! It is also worth noting that you will only gain credit for answers that are at Advanced Higher level. Answers at Higher or National 5 will be ignored and will gain zero marks.

Good luck!

If you have followed the advice given in this introduction you will be well prepared for the Advanced Higher exam. When you sit the exam, take your time and use the experience as an opportunity to show the examiner how much you know. And good luck!

ADVANCED HIGHER FOR CfE

2015 Specimen Question Paper

National Qualifications
SPECIMEN ONLY

SQ05/AH/02

Chemistry
Section 1 — Questions

Date — Not applicable

Duration — 2 hours 30 minutes

Instructions for the completion of Section 1 are given on *Page two* of your question and answer booklet SQ05/AH/01.

Record your answers on the answer grid on *Page three* of your question and answer booklet.

Reference may be made to the Chemistry Higher and Advanced Higher Data Booklet.

Before leaving the examination room you must give your question and answer booklet to the Invigilator; if you do not you may lose all the marks for this paper.

SECTION 1 — 30 marks

Attempt ALL questions

1. An element X forms an ion, X^{3+}, which contains 55 electrons.

   In which block of the Periodic Table would element X be found?

   A   s

   B   p

   C   d

   D   f

2. Which one of the following metal salts will emit radiation of the highest frequency when placed in a Bunsen flame?

   A   Copper(II) chloride

   B   Potassium chloride

   C   Barium chloride

   D   Lithium chloride

3. The following diagram represents a square-planar structure.

   Where ◢━━ and ◿━ represent

   bonding electron pairs

   and ⦿ represents a non-bonding electron pair

   (lone pair).

   Which of the following species could have the structure shown above?

   A SF_4

   B NH_4^+

   C XeF_4

   D AlH_4^-

4. A complex ion with the name hexaamminenickel(II) will have the formula

   A $[Ni(NH_2)_6]^{2+}$

   B $[Ni(NH_3)_6]^{2+}$

   C $[Ni(NH_3)_6]^{4-}$

   D $[Ni(NH_4)_6]^{2+}$.

5. The reaction

$$CO(g) + 3H_2(g) \rightleftharpoons CH_4(g) + H_2O(g)$$

has an equilibrium constant of 3·9 at 950 °C.

The equilibrium concentrations of $CO(g)$, $H_2(g)$ and $H_2O(g)$ at 950 °C are given in the table.

Substance	Equilibrium concentration/mol l^{-1}
$CO(g)$	$5 \cdot 0 \times 10^{-2}$
$H_2(g)$	$1 \cdot 0 \times 10^{-2}$
$H_2O(g)$	$4 \cdot 0 \times 10^{-3}$

What is the equilibrium concentration of $CH_4(g)$, in mol l^{-1}, at 950 °C?

   A $4 \cdot 9 \times 10^{-1}$

   B $3 \cdot 1 \times 10^{-5}$

   C $4 \cdot 9 \times 10^{-5}$

   D $2 \cdot 0 \times 10^{-7}$

6. Which of the following decreases when an aqueous solution of ethanoic acid is diluted?

   A   pH

   B $[H^+]$

   C pK_a

   D K_a

7. The pH of a buffer solution prepared by mixing equal volumes of 0·1 mol l^{-1} ethanoic acid and 0·2 mol l^{-1} sodium ethanoate is

   A   2·1

   B   2·7

   C   4·5

   D   5·1.

8. The graph below shows the pH changes when $0\cdot1\,mol\,l^{-1}$ ammonia solution is added to $50\,cm^3$ of $0\cdot1\,mol\,l^{-1}$ hydrochloric acid solution.

Volume of $0\cdot1\,mol\,l^{-1}\,NH_3$ added/cm^3

Which line in the table shows an indicator which is **not** suitable for use in determining the equivalence point for the above reaction?

	Indicator	pH range of indicator
A	methyl orange	$3\cdot1 - 4\cdot4$
B	bromophenol red	$5\cdot2 - 6\cdot8$
C	bromothymol blue	$6\cdot0 - 7\cdot6$
D	phenolphthalein	$8\cdot3 - 10\cdot0$

9. Ethanoic acid is a weak acid and hydrochloric acid is a strong acid. Which of the following is **not** true?

A   The pH of $0\cdot1\,mol\,l^{-1}$ hydrochloric acid is 1.

B $20\cdot0\,cm^3$ of $0\cdot1\,mol\,l^{-1}$ sodium hydroxide is exactly neutralised by $20\cdot0\,cm^3$ of $0\cdot1\,mol\,l^{-1}$ ethanoic acid.

C   The pH of $0\cdot1\,mol\,l^{-1}$ hydrochloric acid is lower than that of $0\cdot1\,mol\,l^{-1}$ ethanoic acid.

D   The K_a value for ethanoic acid is greater than that of hydrochloric acid.

10. The standard enthalpy of formation of strontium chloride is the enthalpy change for which of the following reactions?

A $Sr(g) + Cl_2(g) \rightarrow SrCl_2(s)$

B $Sr(s) + Cl_2(g) \rightarrow SrCl_2(s)$

C $Sr^{2+}(g) + 2Cl^-(g) \rightarrow SrCl_2(s)$

D $Sr^{2+}(aq) + 2Cl^-(aq) \rightarrow SrCl_2(s)$

11. Which of the following alcohols would have the greatest entropy at 90 °C?

    A   Propan-1-ol

    B   Butan-1-ol

    C   Propan-2-ol

    D   Butan-2-ol

12. For any liquid, $\Delta S_{vaporisation} = \dfrac{\Delta H_{vaporisation}}{T_b}$

    where T_b = boiling point of that liquid.

    For many liquids, $\Delta S_{vaporisation} = 88\,J\,K^{-1}\,mol^{-1}$.

    Assuming that this value is true for water and that its
    $\Delta H_{vaporisation} = 40{\cdot}6\,kJ\,mol^{-1}$, then the boiling point of water is calculated as

    A   0·46 K

    B   2·17 K

    C   373 K

    D   461 K.

13. The order of a reactant in a reaction

    A   can only be obtained by experiment

    B   determines the speed of the overall reaction

    C   is determined by the stoichiometry involved

    D   is the sequence of steps in the reaction mechanism.

14. A suggested mechanism for the reaction

    $2X + Y \rightarrow X_2Y$

    is a two-step process

    $X + Y \rightarrow XY$ (slow)
    $XY + X \rightarrow X_2Y$ (fast)

    This mechanism is consistent with which of the following rate equations?

    A   Rate = k[XY]

    B   Rate = k[X][Y]

    C   Rate = k[X]^2[Y]

    D   Rate = k[X][XY]

15. What volume of $0.2 \, mol \, l^{-1}$ potassium sulfate is required to make, by dilution with water, one litre of a solution with a **potassium** ion concentration of $0.1 \, mol \, l^{-1}$?

   A $500 \, cm^3$

   B $400 \, cm^3$

   C $250 \, cm^3$

   D $100 \, cm^3$

16. The end-on overlap of two atomic orbitals lying along the axis of a bond leads to

   A   hydridisation

   B   a sigma bond

   C   a pi bond

   D   a double bond.

17. Which of the following has nucleophilic properties?

   A   Na

   B Br^+

   C CH_3^+

   D NH_3

18. Carbonyl groups in aldehydes and ketones react with HCN and the product can then be hydrolysed forming a 2-hydroxycarboxylic acid as shown in the equation below.

   When the final product is 2-hydroxy-2-methylbutanoic acid, the starting carbonyl compound is

   A   propanol

   B   propanone

   C   butanal

   D   butanone.

19. Which of the following is a tertiary haloalkane?

    A $CHCl_3$

    B $(CH_3)_3CCl$

    C $(CH_2Cl)_3CH$

    D $(CH_3)_3CCH_2Cl$

20. The structures of three alcohols, P, Q and R are shown.

Which line in the table describes the trends in boiling points and viscosities on moving from P to Q to R?

	Boiling point	Viscosity
A	increases	increases
B	increases	decreases
C	decreases	increases
D	decreases	decreases

21. The Williamson synthesis for the preparation of unsymmetrical ethers (ROR') starting with an alcohol and a haloalkane is summarised in the general equations:

    Step 1: $ROH + Na \rightarrow RO^-Na^+ + \frac{1}{2}H_2$
    Step 2: $RO^-Na^+ + R' X \rightarrow ROR' + Na^+X^-$

    Using propan-2-ol and 2-chlorobutane, the unsymmetrical ether formed would be

    A $CH_3CH_2CH_2OCH(CH_3)CH_2CH_3$

    B $CH_3CH_2CH_2OCH_2CH_2CH_2CH_3$

    C $CH_3CH(CH_3)OCH_2CH_2CH_2CH_3$

    D $CH_3CH(CH_3)OCH(CH_3)CH_2CH_3$.

22.

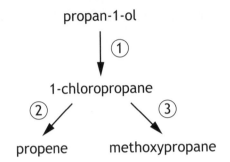

propan-1-ol

1-chloropropane

propene   methoxypropane

Which line in the table is correct for the types of reaction taking place at (1), (2) and (3)?

	Reaction (1)	Reaction (2)	Reaction (3)
A	substitution	elimination	substitution
B	substitution	reduction	substitution
C	addition	reduction	condensation
D	addition	elimination	substitution

23. P $\bigcirc$—Cl

Q $CH_2=CHCl$

R $CH_2=CHCH_2Cl$

Which of the above molecules is/are planar?

A   P only

B   Q and R only

C   P and Q only

D   P, Q and R

24. One mole of which of the following compounds will react with the largest volume of 1 mol l^{-1} hydrochloric acid?

A

B $HOOCCH_2NH_2$

C $H_2NCH_2NH_2$

D CH_3NHCH_3

25. Which of the following amines shows no infra-red absorption between 3300 cm^{-1} and 3500 cm^{-1}?

A $(CH_3)_3N$

B CH_3NHCH_3

C $H_2NCH_2NH_2$

D   (benzene ring)—NH_2

26. Which of the following reactions is **least** likely to take place?

A   (benzene) $\xrightarrow{Br_2/AlCl_3}$ (benzene)—Br

B   (benzene) $\xrightarrow{Br_2/light}$ (benzene)—Br

C   (benzene)—CH_3 $\xrightarrow{Br_2/light}$ (benzene)—CH_2Br

D   (benzene) $\xrightarrow{H_2SO_4/SO_3}$ (benzene)—SO_3H

27. Which of the following analytical techniques would be most suitable to determine quantitatively the concentration of sodium ions in a urine sample?

    A   Mass spectrometry

    B   Infra-red spectroscopy

    C   Atomic emission spectroscopy

    D   Proton nuclear magnetic resonance spectroscopy

28. The hydrolysis of the haloalkane $(CH_3)_3CBr$ was found to take place by an S_N1 mechanism.

    The rate determining step involved the formation of

29. One of the splitting patterns seen in the high resolution ^1H NMR spectrum of 3-methylbutan-1-ol below is shown.

3-methylbutan-1-ol

Which of the following H atoms, H_A, H_B, H_C and H_D, would produce this splitting pattern?

A H_A

B H_B

C H_C

D H_D

30. The table shows the structural formulae of some sulfonamides and their antibacterial activity.

Sulfonamide	Antibacterial activity
	Active
	Active
	Inactive
	Inactive

Which of the following would be an active antibacterial agent?

[END OF SECTION 1. NOW ATTEMPT THE QUESTIONS IN SECTION 2 OF YOUR QUESTION AND ANSWER BOOKLET.]

FOR OFFICIAL USE

National
Qualifications
SPECIMEN ONLY

Mark

SQ05/AH/01

**Chemistry
Section 1 — Answer Grid
and Section 2**

Date — Not applicable

Duration — 2 hours 30 minutes

Fill in these boxes and read what is printed below.

Full name of centre

Town

Forename(s)

Surname

Number of seat

Date of birth

Day	Month	Year	Scottish candidate number
D D	M M	Y Y	

Reference may be made to the Chemistry Higher and Advanced Higher Data Booklet.

Total marks — 100

SECTION 1 — 30 marks

Attempt ALL questions.

Instructions for completion of Section 1 are given on *Page two*.

SECTION 2 — 70 marks

Attempt ALL questions

Write your answers clearly in the spaces provided in this booklet. Additional space for answers and rough work is provided at the end of this booklet. If you use this space you must clearly identify the question number you are attempting. Any rough work must be written in this booklet. You should score through your rough work when you have written your final copy.

Use **blue** or **black** ink.

Before leaving the examination room you must give this booklet to the Invigilator; if you do not, you may lose all the marks for this paper.

SECTION 1— 30 marks

The questions for Section 1 are contained in the question paper SQ05/AH/02.
Read these and record your answers on the answer grid on *Page three* opposite.
Do NOT use gel pens.

1. The answer to each question is **either** A, B, C or D. Decide what your answer is, then fill in the appropriate bubble (see sample question below).

2. There is **only one correct** answer to each question.

3. Any rough working should be done on the additional space for answers and rough work at the end of this booklet.

Sample Question

To show that the ink in a ball-pen consists of a mixture of dyes, the method of separation would be:

    A   fractional distillation

    B   chromatography

    C   fractional crystallisation

    D   filtration.

The correct answer is **B**—chromatography. The answer **B** bubble has been clearly filled in (see below).

Changing an answer

If you decide to change your answer, cancel your first answer by putting a cross through it (see below) and fill in the answer you want. The answer below has been changed to **D**.

If you then decide to change back to an answer you have already scored out, put a tick (✓) to the **right** of the answer you want, as shown below:

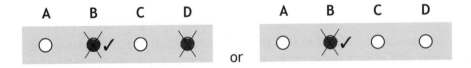
or

SECTION 1 — Answer Grid

	A	B	C	D
1	○	○	○	○
2	○	○	○	○
3	○	○	○	○
4	○	○	○	○
5	○	○	○	○
6	○	○	○	○
7	○	○	○	○
8	○	○	○	○
9	○	○	○	○
10	○	○	○	○
11	○	○	○	○
12	○	○	○	○
13	○	○	○	○
14	○	○	○	○
15	○	○	○	○
16	○	○	○	○
17	○	○	○	○
18	○	○	○	○
19	○	○	○	○
20	○	○	○	○
21	○	○	○	○
22	○	○	○	○
23	○	○	○	○
24	○	○	○	○
25	○	○	○	○
26	○	○	○	○
27	○	○	○	○
28	○	○	○	○
29	○	○	○	○
30	○	○	○	○

MARKS | DO NOT WRITE IN THIS MARGIN

SECTION 2 — 70 marks

Attempt ALL questions

1. In 2002, astronomers observed a red giant star flash 10 000 times brighter than normal. Its electromagnetic spectrum revealed an intense crimson red line, wavelength 670·7 nm.

   (a) Identify an element present in the red giant star that could be responsible for this intense crimson red line in the emission spectrum. **1**

   (b) Explain how the line of red light is produced. **2**

   (c) Calculate the energy, in kJ mol^{-1}, associated with this wavelength. **2**

MARKS

2. Most commercial bleaches contain hypochlorous acid. This acid dissociates as follows:

$$HClO(aq) \quad + \quad H_2O(\ell) \quad \rightleftharpoons \quad H_3O^+(aq) \quad + \quad ClO^-(aq)$$

(a) Identify the conjugate base of hypochlorous acid.

1

(b) Write the expression for the dissociation constant, K_a, for hypochlorous acid.

1

(c) A solution of hypochlorous acid was titrated with sodium hydroxide solution.

The solution at the end point was alkaline.

Explain why the solution at the end point was alkaline.

2

MARKS | DO NOT WRITE IN THIS MARGIN

3. β-carotene is an orange substance and is one of the compounds responsible for the colour of carrots and autumn leaves.

β-carotene

Using your knowledge of chemistry, comment on why compounds, such as β-carotene, are the colour they are.

3

MARKS

4. The ore pyrolusite, MnO_2, was used 30 000 years ago as a black pigment in the cave paintings of Lascaux, France.

   The best known oxide of manganese is possibly potassium permanganate, $KMnO_4$, first made in 1740 for the glass industry. It now has many uses including disinfectants and the removal of organic impurities from waste gases and effluent water.

   (a) (i) State the oxidation number of manganese in MnO_2.   1

        (ii) Using orbital box notation, write the electronic configuration for a manganese ion in MnO_2.   1

        (iii) Explain how your answer is consistent with Hund's rule.   1

   (b) (i) The d orbitals in an isolated manganese atom are degenerate. State the meaning of the term degenerate.   1

MARKS | DO NOT WRITE IN THIS MARGIN

4. (b) (continued)

(ii) The second quantum number, ℓ, is related to the shape of the orbitals.

Draw the shape of an orbital when $\ell = 1$.   **1**

(c) Small amounts of manganese are added to the aluminium used for drinks cans to improve their corrosion resistance. The technique of colorimetry can be used to determine the quantity of manganese in these alloys and involves converting the manganese to permanganate ions, MnO_4^-.

(i) Describe how the technique of colorimetry can be used to determine the concentration of permanganate ions.   **3**

(ii) During colorimetric analysis, $0{\cdot}35\,g$ of an aluminium alloy was dissolved in nitric acid. The manganese in the resulting solution was oxidised and the solution was made up to $250\,cm^3$.

The concentration of this solution was found to be $4{\cdot}25 \times 10^{-4}\,mol\,l^{-1}$.

Calculate the percentage, by mass, of manganese in the alloy.   **2**

MARKS | DO NOT WRITE IN THIS MARGIN

5. A student was investigating the percentage calcium carbonate content in different types of egg shells. The egg shells were ground and approximately 0·4 g were weighed accurately. The shells were placed in a beaker and 20·0 cm³ of 1·00 mol l⁻¹ hydrochloric acid was added. Once the reaction was complete, the solution was made up to 100·0 cm³ in a standard flask.

$$CaCO_3(s) + 2HCl(aq) \rightarrow CaCl_2(aq) + CO_2(g) + H_2O(\ell)$$

(a) State what is meant by weighing accurately, approximately 0·4 g.   **1**

(b) Describe the steps required to prepare the 100·0 cm³ solution.   **2**

(c) 10·0 cm³ aliquots of the solution were titrated against 0·100 mol l⁻¹ standardised sodium hydroxide solution using phenolphthalein as an indicator until concordant results were obtained.

$$NaOH(aq) + HCl(aq) \rightarrow NaCl(aq) + H_2O(\ell)$$

(i) State why the sodium hydroxide solution had to be standardised.   **1**

MARKS | DO NOT WRITE IN THIS MARGIN

5. (c) (continued)

(ii) An egg shell sample of 0·390 g was used in this experiment.

This led to an average titre volume of 12·65 cm³.

Calculate the percentage, by mass, of calcium carbonate present in the egg shell.

4

MARKS | DO NOT WRITE IN THIS MARGIN

6. Omeprazole is a drug commonly used to prevent stomach ulcers. It is described as a proton pump inhibitor as it reduces the ability of enzymes to produce gastric acid. Omeprazole exhibits optical isomerism and is sold as a mixture of both enantiomers.

omeprazole

(a) Write the molecular formula for omeprazole.

1

(b) State the name given to a mixture containing equal amounts of both enantiomers.

1

(c) Suggest the drug classification that best describes omeprazole.

1

(d) Only one omeprazole enantiomer, known as esomeprazole, is active. However, in acidic environments the other, inactive, enantiomer is converted into the active one.

   (i) Explain the benefit of selling the drug as an equal mixture of both enantiomers.

1

MARKS | DO NOT WRITE IN THIS MARGIN

6.   (d)   (continued)

(ii)   The first stage of the conversion of the inactive enantiomer involves the reaction with H^+ ions as shown.

esomeprazole   compound X

Identify a functional group that is present in compound X but not in esomeprazole.

1

7. To make nuclear fuel from uranium ore, the element uranium has to be extracted from the ore before being made into fuel pellets.

   One of the reactions in the production of nuclear fuel from uranium ore is

   $$UO_2(s) \ + \ 4HF(g) \ \rightarrow \ UF_4(s) \ + \ 2H_2O(g) \qquad \Delta H^\circ \ = \ -244 \, kJ \, mol^{-1}$$

   The data in the table below refers to the substances at 298 K.

Substance	$S^\circ / J \, K^{-1} \, mol^{-1}$
$UO_2(s)$	77
$HF(g)$	174
$UF_4(s)$	152
$H_2O(g)$	189

   (a) Use the data to calculate the entropy change, in $J \, K^{-1} \, mol^{-1}$, at 298 K for this reaction.

       2

   (b) Determine, by calculation, whether this reaction is feasible at 298 K.   3

MARKS | DO NOT WRITE IN THIS MARGIN

8. Benzocaine is used to relieve pain and itching caused by conditions such as sunburn, insect bites or stings.

benzocaine

A student was carrying out a project to synthesise benzocaine. Part of the procedure to isolate the synthesised benzocaine is given below.

---

1. Add $20\,cm^3$ of diethyl ether to the reaction mixture and pour into a separating funnel.
2. Add $20\,cm^3$ of distilled water to the separating funnel.
3. Stopper the funnel, invert and gently shake.
4. Allow the aqueous layer to settle to the bottom.

(a) (i) Name the technique described in this part of the procedure.   1

(ii) Outline the next steps the student will need to carry out in order to obtain a maximum yield of solid benzocaine from the reaction mixture.   3

(iii) State **two** properties the solvent must have for it to be appropriate for use in this procedure.   2

MARKS | DO NOT WRITE IN THIS MARGIN

8. **(continued)**

(b) Suggest a second technique that could be used to purify a solid sample of benzocaine.

1

(c) Thin layer chromatography (TLC) was used to help confirm the identity of the product. A sample of product was dissolved in a small volume of solvent and spotted onto a TLC plate. The plate was allowed to develop and the following chromatogram was obtained.

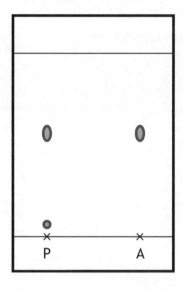

P = student's product

(i) State the name of the substance spotted at A on the TLC plate.

1

(ii) Evaluate the purity of the student's product.

1

MARKS | DO NOT WRITE IN THIS MARGIN

9. At 1000 °C, nitric oxide can combine with hydrogen.

$$2NO(g) \ + \ 2H_2(g) \ \rightarrow \ N_2(g) \ + \ 2H_2O(g)$$

The rate of the above reaction was monitored at different concentrations of NO(g) and H_2(g). The results are shown in the table.

Experiment	[NO] mol l^{-1}	[H$_2$] mol l^{-1}	Initital rate/mol l^{-1} s^{-1}
1	$4 \cdot 00 \times 10^{-3}$	$1 \cdot 00 \times 10^{-3}$	$1 \cdot 20 \times 10^{-5}$
2	$8 \cdot 00 \times 10^{-3}$	$1 \cdot 00 \times 10^{-3}$	$4 \cdot 80 \times 10^{-5}$
3	$8 \cdot 00 \times 10^{-3}$	$4 \cdot 00 \times 10^{-3}$	$1 \cdot 92 \times 10^{-4}$

(a) Determine the order of the reaction with respect to:

    (i) NO(g) 1

    (ii) H$_2$(g). 1

(b) Write the overall rate equation for the reaction. 1

(c) Calculate a value for the rate constant, k, including the appropriate units. 2

MARKS | DO NOT WRITE IN THIS MARGIN

10. Pulegone, first isolated in 1891 from oil of pennyroyal, is a naturally occurring colourless oily liquid with an odour of peppermint. It is classified as a monoterpene and is used in perfumes and aromatherapy.

pulegone

(a) Circle the chiral centre on the structure of the pulegone above.

1

(b) Suggest the most appropriate spectroscopic method for identifying the carbonyl group in pulegone.

1

(c) If pulegone is treated with acid it is converted into two ketones, C_3H_6O and $C_7H_{12}O$.

Draw a possible structure for each of the ketones.

2

MARKS | DO NOT WRITE IN THIS MARGIN

11. Cisplatin was the first member of a class of platinum-containing anti-cancer drugs.

cisplatin

Clinical use of the drug is now limited since cancer cells can develop resistance to it.

(a) (i) Explain the meaning of *cis* in cisplatin.   **1**

(ii) In the cisplatin complex, chloride ions and ammonia molecules are both classed as monodentate ligands.

Explain the term *monodentate*.   **1**

(b) A new drug being trialled, asplatin, is proving to be 10 times more effective than cisplatin.

asplatin

11. (b) (continued)

Asplatin is synthesised by reacting oxoplatin with acetylsalicylic anhydride.

oxoplatin
GFM = 334·1 g

acetylsalicylic anhydride
GFM = 342·3 g

asplatin
GFM = 484·1 g

During a trial synthesis, 5·00 g of oxoplatin was reacted with excess acetylsalicylic anhydride to produce 6·36 g of asplatin.

Calculate the percentage yield.

2

MARKS | DO NOT WRITE IN THIS MARGIN

12. Propanoic acid is commonly used in the food industry as a preservative as it can inhibit the growth of mould and bacteria.

Propanoic acid can be prepared from a variety of small molecules. In industry, the main method of production of propanoic acid is by the reaction of ethene with water and carbon monoxide.

$$C_2H_4 + H_2O + CO \rightarrow CH_3CH_2COOH$$

Using your knowledge of chemistry, outline the possible steps in the synthesis of propanoic acid from small molecules such as ethene or ethanol.

3

MARKS | DO NOT WRITE IN THIS MARGIN

13. Both spectroscopic and chemical analysis can be used to determine the identity of an unknown compound.

The following spectra and chemical data were obtained for a colourless liquid with a pleasant smell.

^{1}H NMR Spectrum

MARKS | DO NOT WRITE IN THIS MARGIN

13. (continued)

Mass Spectrum

Infra-Red Spectrum

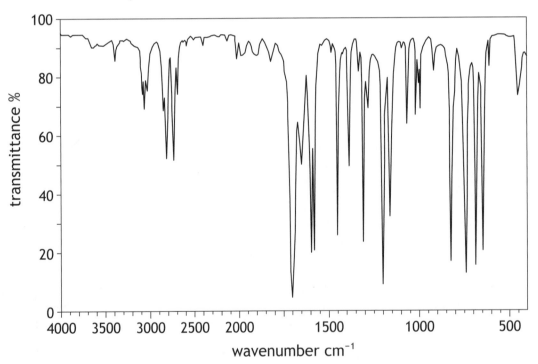

Chemical Data

Composition: C 79·25%; H 5·66%; O 15·09%
Fehling's/Benedict's: brick red precipitate forms
Bromine solution: no change
Aqueous sodium carbonate: no reaction

MARKS

13. **(continued)**

    (a)   (i) Determine the empirical formula for the compound.   2

         (ii) Using the mass spectrum, determine the molecular formula for the compound.   1

    (b) Identify the functional group in the compound which is responsible for the peak at $1700\,cm^{-1}$ in the infra-red spectrum.   1

    (c) Using all of the information, draw a structural formula for the compound.   1

[END OF SPECIMEN QUESTION PAPER]

ADDITIONAL SPACE FOR ANSWERS AND ROUGH WORK

ADDITIONAL SPACE FOR ANSWERS AND ROUGH WORK

[BLANK PAGE]

DO NOT WRITE ON THIS PAGE

ADVANCED HIGHER FOR CfE

Model Paper 1

Whilst this Model Paper has been specially commissioned by Hodder Gibson for use as practice for the Advanced Higher (for Curriculum for Excellence) exams, the key reference document remains the SQA Specimen Paper 2015.

HODDER
GIBSON
LEARN MORE

National
Qualifications
MODEL PAPER 1

Chemistry
Section 1 — Questions

Duration — 2 hours 30 minutes

Instructions for the completion of Section 1 are given on *Page two* of your question and answer booklet.

Record your answers on the answer grid on *Page three* of your question and answer booklet.

Reference may be made to the Chemistry Higher and Advanced Higher Data Booklet.

Before leaving the examination room you must give your question and answer booklet to the Invigilator; if you do not you may lose all the marks for this paper.

SECTION 1 — 30 marks

Attempt ALL questions

1. The diagram shows one of the series of lines in the hydrogen emission spectrum.

Each line

A   represents an energy level within a hydrogen atom

B   results from an electron moving to a higher energy level

C   lies within the visible part of the electromagnetic spectrum

D   results from an excited electron dropping to a lower energy level.

2. An atom has electronic configuration

$1s^2\ 2s^2\ 2p^6\ 3s^2\ 3p^3$.

When ionised, this is likely to form a particle with a charge of

A   5+

B   5−

C   3+

D   3−.

3. Which line in the table correctly describes the arrangement of electron pairs in PF_3 and PF_5?

	PF_3	PF_5
A	Trigonal planar	Trigonal bipyramidal
B	Tetrahedral	Octahedral
C	Tetrahedral	Trigonal bipyramidal
D	Trigonal planar	Octahedral

4. The number of unpaired electrons in a vanadium atom is

    A   0

    B   2

    C   3

    D   5.

5. At a particular temperature, 8·0 mole of NO_2 was placed in a 1 litre container and the NO_2 dissociated by the following reaction:

    $$2NO_2(g) \rightleftharpoons 2NO(g) + O_2(g)$$

    At equilibrium the concentration of NO(g) is 2·0 mol l^{-1}.

    The equilibrium constant will have a value of

    A   0·11

    B   0·22

    C   0·33

    D   9·00.

6. The pH of 0·025 mol l^{-1} H_2SO_4(aq) is

    A   1·3

    B   −1·6

    C   1·6

    D   1·9.

7. Ammonia, NH_3, can react with hydrogen chloride gas to form the salt ammonium chloride.

    $NH_3(g) + HCl(g) \rightarrow NH_4Cl(s)$

    In this reaction, ammonia is acting as

    A   a proton donor

    B   an acid

    C   a proton acceptor

    D   a conjugate base.

8. Which of the following would not act as a buffer solution?

   A   Ethanoic acid and potassium ethanoate

   B   Nitric acid and potassium nitrate

   C   Ammonia and ammonium nitrate

   D   Hexanoic acid and magnesium hexanoate

9. Which of the following indicators should be used in the titration of methanoic acid with sodium hydroxide?

   A   Congo red, pH range 3·0–5·0

   B   Methyl red, pH range 4·2–6·2

   C   Phenolphthalein, pH range 8·0–9·8

   D   Bromothymol blue, pH range 6·0–7·6

10. The equation representing the standard enthalpy of formation of calcium carbonate is

    A $Ca(s) + C(s) + 1\frac{1}{2}O_2(g) \rightarrow CaCO_3(s)$

    B $Ca(g) + C(g) + 1\frac{1}{2}O_2(g) \rightarrow CaCO_3(g)$

    C $Ca(s) + CO_2(g) + \frac{1}{2}O_2(g) \rightarrow CaCO_3(s)$

    D $Ca(s) + C(s) + 3O(g) \rightarrow 2CaCO_3(s)$

11. Which of the following reactions will have a positive ΔS° value?

    A $C_2H_4(g) + H_2(g) \rightarrow C_2H_6(g)$

    B $2K(s) + 2H_2O(l) \rightarrow 2KOH(aq) + H_2(g)$

    C $Pb(NO_3)_2(aq) + 2KI(aq) \rightarrow PbI_2(s) + 2KNO_3(aq)$

    D $2H_2(g) + O_2(g) \rightarrow 2H_2O(l)$

12. The ΔG° at 298 K for a chemical reaction with $\Delta H^\circ = 10\,kJ\,mol^{-1}$ and $\Delta S^\circ = -14\,J\,K^{-1}\,mol^{-1}$ is

    A $14\,kJ\,mol^{-1}$

    B $-14\,kJ\,mol^{-1}$

    C $-4162\,kJ\,mol^{-1}$

    D $4162\,kJ\,mol^{-1}$.

13. The following data refer to initial reaction rates obtained for the reaction

X + Y + Z → products

Run	Relative concentrations			Relative initial rate
	[X]	[Y]	[Z]	
1	1·0	1·0	1·0	0·3
2	1·0	2·0	1·0	0·6
3	2·0	2·0	1·0	1·2
4	2·0	1·0	2·0	0·6

These data fit the rate equation

A   Rate = k[X]

B   Rate = k[X][Y]

C   Rate = k[X][Y]^2

D   Rate = k[X][Y][Z].

14.

H   H
 \ |
  C   C   H
 / \\ / \\ /
H   C   C
     | |
     H   H

The correct number of sigma and pi bonds in the alkene shown above is

A   nine sigma and two pi

B   seven sigma and four pi

C   eleven sigma and two pi

D   eleven sigma and four pi.

15. The sideways overlap of the two p orbitals shown below will result in the formation of a

A   sigma bond

B   pi bond

C   hybrid orbital

D   sigma bond and a pi bond.

16. Which of the following has a geometric isomer?

A	H—C—C—H with H, H above and Cl, Cl below
B	C=C with H and H, Cl and Cl
C	C=C with Cl and Cl, Cl and Cl
D	H—C—C—H with Cl, Cl above and Cl, Cl below

17. Propene can react with hydrogen chloride to form 2-chloropropane. Which line in the table correctly describes this reaction?

	Propene is acting as	Type of bond fission
A	a nucleophile	heterolytic
B	an electrophile	homolytic
C	a nucleophile	homolytic
D	an electrophile	heterolytic

18. Alkenes react with ozone (O_3) to form ozonides which can be hydrolysed to give carbonyl compounds.

$$\underset{/}{\overset{\backslash}{}}C{=}C\underset{\backslash}{\overset{/}{}} \xrightarrow{\;O_3\;} \text{OZONIDE} \xrightarrow{\;\text{hydrolysis}\;} \underset{/}{\overset{\backslash}{}}C{=}O + O{=}C\underset{\backslash}{\overset{/}{}}$$

Which of the following alkenes will produce a mixture of propanone and ethanal when acted upon in this way?

A $CH_3CH{=}CHCH_2CH_3$

B $CH_3CH{=}CHCH_3$

C $(CH_3)_2C{=}CH_2$

D $CH_3CH{=}C(CH_3)_2$

19. Which line in the table below correctly describes the reaction of ethanal with LiAlH4?

	$LiAlH_4$ acts as	Product formed
A	a reducing agent	ethanoic acid
B	an oxidising agent	ethanol
C	a reducing agent	ethanol
D	an oxidising agent	ethanoic acid

20. Which of the following compounds would produce ethyl propanoate when reacted with ethanol?

A	$CH_3CH_2-\overset{\overset{\textstyle O}{\|}}{C}-H$
B	$CH_3CH_2-\overset{\overset{\textstyle O}{\|}}{C}-Cl$
C	$CH_3-\overset{\overset{\textstyle O}{\|}}{C}-CH_3$
D	$CH_3-\overset{\overset{\textstyle OH}{\|}}{C}H-CH_3$

21. Which line in the table correctly describes the reaction of benzene with a mixture of concentrated nitric acid and concentrated sulfuric acid?

	Type of reaction	Product formed
A	electrophilic substitution	NO_2 (on benzene ring)
B	electrophilic substitution	SO_3H (on benzene ring)
C	nucleophilic substitution	NO_2 (on benzene ring)
D	nucleophilic substitution	SO_3H (on benzene ring)

22. Which of the following cannot be applied to benzene?

A sp^2 hybridised carbon atoms

B   Takes part in addition reactions

C   p orbitals overlap to produce a delocalised ring of electrons

D   Planar molecule

23.

The reagents required for the above reaction are

A CH_3CH_2Cl

B CH_3CH_3 and UV light

C CH_3CH_2Cl and $AlCl_3$

D $CH_3CH_2^-$.

24.   Which of the following compounds is most likely to show peaks on a mass spectrum at m/z 45 and m/z 31?

A   Ethanal

B   Ethane

C   Ethanoic acid

D   Ethanol

25.   Which of the following spectroscopic techniques involves vibration of chemical bonds?

A   Mass spectrometry

B   Infrared spectroscopy

C   Atomic emission spectroscopy

D   Proton nuclear magnetic resonance spectroscopy

26. Which of the following compounds is likely to produce the proton NMR spectrum shown below?

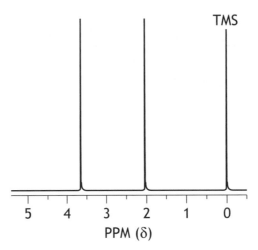

TMS

5   4   3   2   1   0

PPM (δ)

A	H—C—C—H (with H above and below each C)
B	H—C—C—O—H (with H above and below each C)
C	H—C—C(=O)—O—C—H (with H on each C)
D	H—C—C(=O)—H (with H above and below first C, H on carbonyl)

27. Ipratropium is the active ingredient in a medicine used to relieve asthma. The medicine works by binding to a receptor in the lungs causing a specific reaction in the lungs to stop. This allows the patient to breathe more easily.

Which line in the table correctly describes how this medicine works?

	Ipratropium acts as an	Ipratropium binds to
A	agonist	protein
B	antagonist	protein
C	agonist	fat
D	antagonist	fat

28. The most appropriate pieces of equipment to use when diluting a solution by a factor of 10 would be

    A   a $10.0\,cm^3$ pipette and a $100\,cm^3$ measuring cylinder

    B   a $10.0\,cm^3$ pipette and a $50\,cm^3$ standard flask

    C   a $25.0\,cm^3$ measuring cylinder and a $250\,cm^3$ standard flask

    D   a $25.0\,cm^3$ pipette and a $250\,cm^3$ standard flask.

29. Using thin-layer chromatography the components of a mixture can be identified by their R_f values.

    Which of the following statements is **true** about the R_f value of an individual component of a mixture?

    A   The type of stationary phase has no effect on the R_f value.

    B   The polarity of the component has no effect on the R_f value.

    C   The composition of the mobile phase has no effect on the R_f value.

    D   The distance the solvent front moves has no effect on the R_f value.

30. An excess of sodium sulfate was added to a solution of a barium compound to precipitate all the barium ions as barium sulfate, $BaSO_4$. (GFM of $BaSO_4$ = $233.4\,g$.)

    How many grams of barium are in $0.458\,g$ of the barium compound if a solution of this sample gave $0.513\,g$ of $BaSO_4$ precipitate?

    A $0.032\,g$

    B $0.055\,g$

    C $0.269\,g$

    D $0.302\,g$

**[END OF SECTION 1. NOW ATTEMPT THE QUESTIONS IN SECTION 2
OF YOUR QUESTION AND ANSWER BOOKLET.]**

FOR OFFICIAL USE

National
Qualifications
MODEL PAPER 1

Mark

Chemistry
Section 1 — Answer Grid and Section 2

Duration — 2 hours 30 minutes

Fill in these boxes and read what is printed below.

Full name of centre

Town

Forename(s)

Surname

Number of seat

Date of birth

Day   Month   Year

D D   M M   Y Y

Scottish candidate number

Reference may be made to the Chemistry Higher and Advanced Higher Data Booklet.

Total marks — 100

SECTION 1 — 30 marks

Attempt ALL questions.

Instructions for completion of Section 1 are given on *Page two*.

SECTION 2 — 70 marks

Attempt ALL questions

Write your answers clearly in the spaces provided in this booklet. Additional space for answers and rough work is provided at the end of this booklet. If you use this space you must clearly identify the question number you are attempting. Any rough work must be written in this booklet. You should score through your rough work when you have written your final copy.

Use **blue** or **black** ink.

Before leaving the examination room you must give this booklet to the Invigilator; if you do not, you may lose all the marks for this paper.

SECTION 1 — 30 marks

The questions for Section 1 are contained in the question paper on *Page four*.
Read these and record your answers on the answer grid on *Page three* opposite.
Do NOT use gel pens.

1. The answer to each question is **either** A, B, C or D. Decide what your answer is, then fill in the appropriate bubble (see sample question below).

2. There is **only one correct** answer to each question.

3. Any rough working should be done on the additional space for answers and rough work at the end of this booklet.

Sample Question

To show that the ink in a ball-pen consists of a mixture of dyes, the method of separation would be:

    A   fractional distillation

    B   chromatography

    C   fractional crystallisation

    D   filtration.

The correct answer is **B**—chromatography. The answer **B** bubble has been clearly filled in (see below).

Changing an answer

If you decide to change your answer, cancel your first answer by putting a cross through it (see below) and fill in the answer you want. The answer below has been changed to **D**.

If you then decide to change back to an answer you have already scored out, put a tick (✓) to the **right** of the answer you want, as shown below:

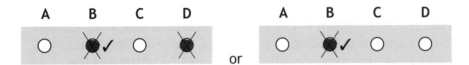

SECTION 1 — Answer Grid

	A	B	C	D
1	○	○	○	○
2	○	○	○	○
3	○	○	○	○
4	○	○	○	○
5	○	○	○	○
6	○	○	○	○
7	○	○	○	○
8	○	○	○	○
9	○	○	○	○
10	○	○	○	○
11	○	○	○	○
12	○	○	○	○
13	○	○	○	○
14	○	○	○	○
15	○	○	○	○
16	○	○	○	○
17	○	○	○	○
18	○	○	○	○
19	○	○	○	○
20	○	○	○	○
21	○	○	○	○
22	○	○	○	○
23	○	○	○	○
24	○	○	○	○
25	○	○	○	○
26	○	○	○	○
27	○	○	○	○
28	○	○	○	○
29	○	○	○	○
30	○	○	○	○

SECTION 2 — 70 marks

Attempt ALL questions

MARKS | DO NOT WRITE IN THIS MARGIN

1. When a high voltage is applied to a lamp filled with neon gas, a line emission spectrum is produced.

    400 nm   500 nm   600 nm   700 nm

    (a) Explain how a line in the line emission spectrum is produced.   2

    (b) Calculate the energy, in $kJ\,mol^{-1}$, associated with the line at 585 nm.   2

    (c) A sample of dry air was found to contain 18 ppm neon. Calculate the volume of neon present in a 10 kg sample of dry air assuming that the molar volume was $22\cdot4\,litres\,mol^{-1}$.   3

MARKS | DO NOT WRITE IN THIS MARGIN

2. (a) A salt of methanoic acid, sodium methanoate is commonly used as a food preservative.

Explain why sodium methanoate solution has a pH greater than 7.

In your answer you should mention the two equilibria involved.   **2**

(b) When sodium m ethanoate is added to methanoic acid, a buffer solution is formed.

0·50 moles of sodium methanoate was added to $200\,cm^3$ of a $0·25\,mol\,l^{-1}$ solution of methanoic acid at 298 K, pK_a = 3·75.

Calculate the pH of the buffer solution formed.   **2**

MARKS | DO NOT WRITE IN THIS MARGIN

3. A student stated that, "Ionisation energy always increases across a period in the periodic table due to the increase in nuclear charge."

Using your knowledge of chemistry, discuss the accuracy of this statement.   3

MARKS | DO NOT WRITE IN THIS MARGIN

4. As part of an investigation, a student was analysing the metallic content of a key known to be composed of a copper/nickel alloy.

The key was dissolved in nitric acid and the resulting solution diluted to $1000\,cm^3$ in a standard flask using tap water. Three $25\cdot0\,cm^3$ samples of the nitrate solution were pipetted into separate conical flasks and approximately $10\,g$ of solid potassium iodide were added. Iodine was produced as shown in the equation.

$$2Cu^{2+}(aq) + 4I^-(aq) \rightarrow 2CuI(s) + I_2(aq)$$

The liberated iodine was titrated against standardised $0\cdot102\,mol\,l^{-1}$ sodium thiosulfate solution. Starch indicator was added near the end point of the titration.

$$I_2(aq) + 2S_2O_3^{2-}(aq) \rightarrow 2I^-(aq) + S_4O_6^{2-}(aq)$$

The results, for the volume of thiosulfate used, are given in the table.

	Titration 1	Titration 2	Titration 3
Final volume/cm³	16·30	31·50	46·80
Initial volume/cm³	0·30	16·30	31·50
Volume added/cm³	16·00	15·20	15·30

(a) From the results calculate the mass of copper in the key.   3

(b) Suggest how the accuracy of the analysis could be improved.   1

MARKS

4. **(continued)**

(c) When copper(II) is dissolved in water, a light blue coloured solution is formed which has the formula $[Cu(H_2O)_6]^{2+}$. Reaction of this solution with excess ammonia results in the formation of the complex $[Cu(NH_3)_4(H_2O)_2]^{2+}$ which has a deep blue colour.

(i) State the name of the complex ion $[Cu(H_2O)_6]^{2+}$.   1

(ii) Draw a structural formula for $[Cu(H_2O)_6]^{2+}$ showing the shape of the complex.   1

(iii) Explain why both these complexes are coloured and explain why changing the ligand from water to ammonia results in a change in colour.   3

MARKS | DO NOT WRITE IN THIS MARGIN

5. Barium carbonate decomposes on heating.

$$BaCO_3(s) \rightarrow BaO(s) + CO_2(g) \quad \Delta H^0 = +266 \, kJ \, mol^{-1}$$

(a) Using the data from the table below, calculate the standard entropy change, ΔS^0, in $J \, K^{-1} \, mol^{-1}$, for the reaction.

Substance	Standard entropy, $S°/J \, K^{-1} \, mol^{-1}$
$BaCO_3(s)$	112·0
$BaO(s)$	72·1
$CO_2(g)$	213·8

2

(b) Calculate the temperature at which the decomposition of barium carbonate just becomes feasible.

3

MARKS

6. The bromate ion, BrO_3^-, is a useful oxidising agent.

(a) Calculate the oxidation number of bromine in the bromate ion.   1

(b) The following table of results was obtained for the reaction between bromate ions and bromide ions under acidic conditions.

$$BrO_3^-(aq) + 5Br^-(aq) + 6H^+(aq) \rightarrow 3Br_2(aq) + 3H_2O(l)$$

Experiment	$[BrO_3^-]/mol\,l^{-1}$	$[Br^-]/mol\,l^{-1}$	$[H^+]/mol\,l^{-1}$	Initial rate/$mol\,l^{-1}s^{-1}$
1	0·05	0·05	0·05	$5·0 \times 10^{-5}$
2	0·10	0·05	0·05	$1·0 \times 10^{-4}$
3	0·10	0·10	0·05	$2·0 \times 10^{-4}$
4	0·05	0·05	0·10	$2·0 \times 10^{-4}$

(i) The order of reaction with respect to BrO_3^- is 1st order. Deduce the order of reaction with respect to Br^- and H^+.   2

(ii) Write the rate equation for the reaction.   1

(iii) Calculate the rate constant for this reaction giving the appropriate units.   2

MARKS | DO NOT WRITE IN THIS MARGIN

7. The electronic spectra of molecules can be described in terms of the wavelength of maximum absorbance, λ_{max}.

The table below shows a number of compounds with their corresponding λ_{max} values.

Compound	λ_{max}/nm
1.	217
2.	227
3.	263
4.	352

(a) Compound 1 is buta-1,3-diene.

Name compound 2.   **1**

(b) Draw the most likely structure for the compound with $\lambda_{max} = 291$ nm.   **1**

(c) The compounds shown have a system of alternating single and double bonds.

What word is used to describe this type of system?   **1**

(d) From the information shown in the table draw a conclusion relating the energy difference between the HOMO and the LUMO as the number of alternating single and double bonds increases.   **1**

MARKS | DO NOT WRITE IN THIS MARGIN

7. **(continued)**

(e) β-carotene, λ_{max} = 452 nm gives the orange colour to carrots and has the structure

whereas α-carotene, λ_{max} = 434 nm is found in oranges and has the structure

Explain why there is a difference in the λ_{max} values for these two structures.

1

(f) The pink colour of cooked salmon and lobster is due to astaxanthin which has the structure

This molecule is optically active.

Draw part of the molecule and circle one of the asymmetric carbon atoms responsible for this optical activity.

1

MARKS | DO NOT WRITE IN THIS MARGIN

8. 2-Methoxy-2-methylpropane is a compound added to unleaded petrol as a 'knock inhibitor'. It can be synthesised by the reaction of methoxide ions with 2-chloro-2-methylpropane.

$$H_3C-\overset{\overset{\displaystyle H_3C}{|}}{\underset{\underset{\displaystyle H_3C}{|}}{C}}-Cl \ + \ H_3C-O^- \longrightarrow H_3C-\overset{\overset{\displaystyle H_3C}{|}}{\underset{\underset{\displaystyle H_3C}{|}}{C}}-O\diagdown_{CH_3}$$

2-chloro-   methoxide ion   2-methoxy-
2-methylpropane   2-methylpropane

(a) To which class of organic compounds does 2-methoxy-2-methylpropane belong?

1

(b) How can the methoxide ion be prepared from methanol?

1

(c) The preparation of the 2-methoxy-2-methylpropane proceeds by a S_N1 mechanism.

   (i) Clearly showing the electron shifts, outline the step(s) involved.

3

   (ii) Suggest why this reaction is more likely to proceed by a S_N1 mechanism rather than a S_N2.

1

(d) 2-Methoxybutane is an isomer of 2-methoxy-2-methylpropane. When 2-methoxybutane is created, a racemic mixture is formed.

   (i) What is a racemic mixture?

1

   (ii) What would happen to polarised light if it was passed through a racemic mixture?

1

MARKS

9. A laboratory technician prepared a solution of potassium iodate, KIO_3(aq), for use in a titration.

   (a) Potassium iodate is an example of a primary standard. State one of the characteristics of a primary standard.

   1

   (b) Describe the steps the technician should take to prepare a standard solution of potassium iodate.

   2

   (c) Describe how $100\,cm^3$ of $0 \cdot 10\,mol\,l^{-1}$ potassium iodate solution would be prepared from a standard $1 \cdot 00\,mol\,l^{-1}$ potassium carbonate solution.

   2

MARKS | DO NOT WRITE IN THIS MARGIN

10. Methyl propanoate can be made by reacting methanol with propanoic acid.

$CH_3OH + CH_3CH_2COOH \rightleftharpoons CH_3OOCCH_2CH_3 + H_2O$

(a) State the type of reaction occurring when methyl propanoate is formed from methanol and propanoic acid.

1

(b) Propanoic acid can be formed from the acid hydrolysis of propane nitrile. Draw a structural formula for propane nitrile.

1

(c) An ^1H NMR spectrum of methyl propanoate is shown below. On the structure, label the H atoms corresponding to Hc.

1

MARKS | DO NOT WRITE IN THIS MARGIN

10.   (continued)

(d) A student suggested that the conversion of methanol and propanoic acid into methyl propanoate could be followed using analytical techniques such as chromatography and infrared spectroscopy allowing a chemist to know when the product ester had formed. Using your knowledge of chemistry, explain how these techniques could be used in this way.

3

MARKS | DO NOT WRITE IN THIS MARGIN

11. Sulfa drugs are compounds with antibiotic properties. Sulfa drugs can be prepared from a solid compound called sulfanilamide.

Sulfanilamide is prepared in a six-stage synthesis. The equation for the final step in the synthesis is shown.

$C_8H_{10}N_2SO_3$
4-acetamidobenzenesulfonamide

$C_6H_8N_2SO_2$
sulfanilamide

(a) Converting the 4-acetamidobenzesulfonamide into sulfanilamide involves gentle heating of the 4-acetamidobenzesulfonamide in a solution of water and hydrochloric acid in a round bottom flask for at least 30 minutes.

Name two pieces of equipment necessary to carry out this reaction successfully.

2

(b) The sulfanilamide is separated from the reaction mixture and recrystallised from boiling water.

(i) Why is recystallisation necessary?

1

(ii) Describe the solubilty of pure sulfanilamide in cold and hot water.

1

(iii) Simple filtration to isolate the pure sulfanilamide is very slow.

How could the filtration be speeded up?

1

MARKS

11. **(continued)**

(c) Calculate the percentage yield of sulfanilamide if 4·282 g of 4-acetamidobenzenesulfonamide produced 2·237 g of sulfanilamide.   **3**

(d) Describe how a mixed melting point experiment would be carried out and the result used to confirm that the product was pure.   **2**

(e) Suggest another analytical technique which could be used to indicate whether the final sample is pure.   **1**

[END OF MODEL PAPER]

ADDITIONAL SPACE FOR ANSWERS AND ROUGH WORK

ADDITIONAL SPACE FOR ANSWERS AND ROUGH WORK

ADVANCED HIGHER FOR CfE

Model Paper 2

Whilst this Model Paper has been specially commissioned by Hodder Gibson for use as practice for the Advanced Higher (for Curriculum for Excellence) exams, the key reference document remains the SQA Specimen Paper 2015.

National
Qualifications
MODEL PAPER 2

**Chemistry
Section 1 — Questions**

Duration — 2 hours 30 minutes

Instructions for the completion of Section 1 are given on *Page two* of your question and answer booklet SQ05/AH/01.

Record your answers on the answer grid on *Page three* of your question and answer booklet.

Reference may be made to the Chemistry Higher and Advanced Higher Data Booklet.

Before leaving the examination room you must give your question and answer booklet to the Invigilator; if you do not you may lose all the marks for this paper.

SECTION 1 — 30 marks

Attempt ALL questions

1. Element X can be ionised to form X^{2+} which has the electronic configuration $1s^2 2s^2 2p^6$.

   In which block of the periodic table would X be found?

   A   s

   B   p

   C   d

   D   f

2. A student observed a line emission spectrum by using a spectroscope to view the yellow colour emitted by a sodium vapour lamp. Which statement best describes the line emission spectrum for sodium?

   The line emission spectrum results from

   A   the emission of photons as electrons drop to lower energy levels

   B   the emission of electrons as they drop from high to low energy levels

   C   electrons absorbing energy to move from lower energy d orbitals to higher energy d orbitals

   D   electrons absorbing energy to move from the LUMO to the HOMO.

3. Which of the following statements for the first row transition metals is always true?

   A   2+ ions are formed by removing electrons from 3d orbitals.

   B   The 3d orbitals remain degenerate when ligands bond to the metal.

   C   All electronic configurations obey the Aufbau principle.

   D   They have at least one electron in the 4s orbital.

4. The orbital box notation shown below for $1s^2$ does not conform to Pauli's exclusion principle.

   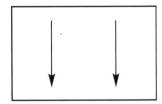

   Which quantum number should be changed to give the correct $1s^2$ orbital box notation?

   A   Principal quantum number

   B   Angular momentum quantum number

   C   Magnetic quantum number

   D   Spin quantum number

5. Which line in the graph represents the trend in successive ionisation energies of a Group 2 element?

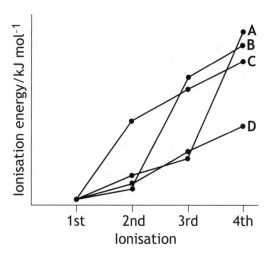

6. A dative covalent bond is present in

A NH_3

B NH_4^+

C CH_3NH_2

D CH_3NH^-.

7. For the following equilibrium

$N_2(g) + 3H_2(g) \rightleftharpoons 2NH_3(g) \; \Delta H = -52 \, kJ \, mol^{-1}$

the value of K could be decreased by

A using an iron catalyst

B increasing the temperature

C increasing the pressure

D carrying out the reaction in an open system.

8. Equal concentrations of the following salt solutions were prepared.

Which of the following would have the highest pH value?

A Potassium methanoate

B Potassium chloride

C Potassium ethanoate

D Potassium benzoate

9. The pH of a buffer solution created from mixing equal volumes of $0.01\,mol\,l^{-1}$ methanoic acid with $0.05\,mol\,l^{-1}$ sodium methanoate would be

    A   3·75

    B   3·05

    C   4·45

    D   2·88.

10. Which of the following will lead to a decrease in entropy?

    A   Water evaporating

    B   Combustion of methanol

    C   Dehydration of ethanol

    D   Hydration of ethene

11. $CaCO_3(s) \rightarrow CaO(s) + CO_2(g)$

    $\Delta H^0 = 178\,kJ\,mol^{-1}$ $\Delta S^0 = 161\,J\,K^{-1}\,mol^{-1}$

    This reaction is thermodynamically feasible

    A   at any temperature

    B   at all temperatures below $904\,K$

    C   at all temperatures above $904\,K$

    D   only at temperatures above $1106\,K$.

12. NO(g) and $Cl_2(g)$ react according to the following equation:

    $2NO(g) + Cl_2(g) \rightarrow 2NOCl(g)$

    Which of the following statements is false if the reaction is 1st order with respect to NO(g) and 1st order with respect to $Cl_2(g)$?

    A   The overall order of this reaction is 2.

    B   The rate equation is rate $= k[NO][Cl_2]$.

    C   The units of k would be $mol\,l^{-1}\,s^{-1}$.

    D   Doubling the concentration of NO would double the reaction rate.

13. A kinetics study of the reaction A + B → C gave the results shown in the table below.

Initial [A]/mol l^{-1}	Initial [B]/mol l^{-1}	Time to form a fixed concentration of C/s
0·2	0·2	24
0·2	0·4	12
0·4	0·2	24

The rate equation for the reaction is

A   rate = k[A][B]^2

B   rate = k[B]^2

C   rate = k[A][B]

D   rate = k[B].

14. Which of the following could be applied to the bonding in propane?

A   Sigma and pi bonds

B   sp^3 and sp^2 hybrid orbitals

C   sp^3 hybrid orbitals and sigma bonds

D   sp^3 hybrid orbitals and p orbital overlap to form pi bonds

15. Which of the following molecules does not contain a conjugated system?

16.

is the skeletal formula for

- A   pentan-2-ol
- B   2-methylbutan-4-ol
- C   2-methylbutan-2-ol
- D   3-methylbutan-2-ol.

17. Which of the following is an example of a tertiary haloalkane?

- A CH_3Cl
- B $(CH_3)_3CCl$
- C $(CH_3)_2CHCl$
- D $(CH_2Cl)_4C$

18. Which of the following statements could not be applied to a racemic mixture?

- A   It is composed of two enantiomers.
- B   It would cause a rotation of polarised light.
- C   It contains two compounds which are non-superimposable mirror images.
- D   It contains optical isomers.

19. Which of the following equations does **not** represent a nucleophilic substitution?

- A $C_3H_7Br + KOH \xrightarrow{ethanol} C_3H_6 + KBr + H_2O$
- B $C_3H_7Br + KCN \xrightarrow{ethanol} C_3H_7CN + KBr$
- C $C_2H_5Cl + C_2H_5ONa \xrightarrow{ethanol} C_2H_5OC_2H_5 + NaCl$
- D $C_2H_5Br + NaOH \xrightarrow{water} C_2H_5OH + NaBr$

20. Which of the following reactions does not involve heterolytic fission?

- A $C_2H_6 + Br_2 \rightarrow C_2H_5Br + HBr$
- B $C_2H_4 + Br_2 \rightarrow C_2H_4Br_2$
- C $C_2H_4 + HBr \rightarrow C_2H_5Br$
- D $CH_3OH + HCOOH \rightarrow CH_3OOCH + H_2O$

21. Which of the following reactions would not produce propanoic acid?

    A   Hydrolysis of propanenitrile

    B   Hydrolysis of methyl propanoate

    C   Oxidation of propan-2-ol

    D   Oxidation of propanal

22. Naphthalene and anthracene are examples of polycyclic aromatic hydrocarbons.

    Naphthalene:

    structural formula

    molecular formula $C_{10}H_8$

    Anthracene:

    structural formula

    The molecular formula of anthracene is

    A $C_{14}H_{10}$

    B $C_{12}H_{10}$

    C $C_{14}H_{12}$

    D $C_{12}H_{12}$.

23. In the reaction between benzene and nitric acid in the presence of concentrated sulfuric acid

    + HNO_3 ⟶ NO$_2$ + H_2O

    A   the benzene molecule acts as an electrophile

    B   the ion, NO_3^-, acts as a nucleophile

    C   the ion, NO_2^+, acts as an electrophile

    D   the HNO_3 is oxidised.

24. Comparing methyl amine, CH_3NH_2, with butyl amine, $C_4H_9NH_2$, methyl amine will have a

    A   higher volatility and a higher solubility in water

    B   higher volatility and a lower solubility in water

    C   lower volatility and a higher solubility in water

    D   lower volatility and a lower solubility in water.

Page seven

25. Elemental analysis of an organic compound showed it contained 70·6% carbon, 23·5% oxygen and 5·9% hydrogen by mass.

The structural formula of the compound could be

A CH_2CHCH_2COOH

B   ⌬—OH
       ⌬—OH

C   ⌬—CH_2CO_2H

D   ⌬—CH_2CHO

26. An organic compound with empirical formula, C_2H_4O, has major peaks at $1715\,cm^{-1}$ and $3300\,cm^{-1}$ in its infrared spectrum.

The structural formula of the compound could be

A CH_3CHO

B CH_3COOH

C $CH_3COOCH_2CH_3$

D $CH_3CH_2CH_2COOH$.

27. Which of the following compounds could have produced the splitting pattern on an ^1H NMR spectrum as shown below?

28. A drug containing a carboxyl group can bind to an amino group on a receptor site in three different ways.

The drug with the following structure

could bind to the same site

A   only by ionic interaction

B   only as a hydrogen-bond donor

C   only as a hydrogen-bond acceptor

D   both as a hydrogen-bond donor and acceptor.

29. The stability of a covalent bond is related to its bond order, which can be defined as follows:

bond order = $\frac{1}{2}$ (number of bonding electrons – number of anti-bonding electrons)

The molecular orbital diagram for oxygen is shown. The anti-bonding orbitals are denoted by *.

The bond order for a molecule of oxygen is

A   0

B   1

C   2

D   3.

30. Which one of the following is not suitable for the preparation of a primary standard in volumetric analysis?

A   Anhydrous sodium carbonate

B   Sodium hydroxide

C   Oxalic acid

D   Potassium iodate

**[END OF SECTION 1. NOW ATTEMPT THE QUESTIONS IN SECTION 2
OF YOUR QUESTION AND ANSWER BOOKLET.]**

National
Qualifications
MODEL PAPER 2

Mark

Chemistry
Section 1 — Answer Grid
and Section 2

Duration — 2 hours 30 minutes

Fill in these boxes and read what is printed below.

Full name of centre

Town

Forename(s)

Surname

Number of seat

Date of birth

Day   Month   Year

D D   M M   Y Y

Scottish candidate number

Reference may be made to the Chemistry Higher and Advanced Higher Data Booklet.

Total marks — 100

SECTION 1 — 30 marks

Attempt ALL questions.

Instructions for completion of Section 1 are given on *Page two*.

SECTION 2 — 70 marks

Attempt ALL questions

Write your answers clearly in the spaces provided in this booklet. Additional space for answers and rough work is provided at the end of this booklet. If you use this space you must clearly identify the question number you are attempting. Any rough work must be written in this booklet. You should score through your rough work when you have written your final copy.

Use **blue** or **black** ink.

Before leaving the examination room you must give this booklet to the Invigilator; if you do not, you may lose all the marks for this paper.

HODDER
GIBSON
LEARN MORE

SECTION 1— 30 marks

The questions for Section 1 are contained in the question paper on *Page Four*.
Read these and record your answers on the answer grid on *Page three* opposite.
Do NOT use gel pens.

1.   The answer to each question is **either** A, B, C or D. Decide what your answer is, then fill in the appropriate bubble (see sample question below).

2.   There is **only one correct** answer to each question.

3.   Any rough working should be done on the additional space for answers and rough work at the end of this booklet.

Sample Question

To show that the ink in a ball-pen consists of a mixture of dyes, the method of separation would be:

    A   fractional distillation

    B   chromatography

    C   fractional crystallisation

    D   filtration.

The correct answer is **B**—chromatography. The answer **B** bubble has been clearly filled in (see below).

Changing an answer

If you decide to change your answer, cancel your first answer by putting a cross through it (see below) and fill in the answer you want. The answer below has been changed to **D**.

If you then decide to change back to an answer you have already scored out, put a tick (✓) to the **right** of the answer you want, as shown below:

SECTION 1 — Answer Grid

	A	B	C	D
1	○	○	○	○
2	○	○	○	○
3	○	○	○	○
4	○	○	○	○
5	○	○	○	○
6	○	○	○	○
7	○	○	○	○
8	○	○	○	○
9	○	○	○	○
10	○	○	○	○
11	○	○	○	○
12	○	○	○	○
13	○	○	○	○
14	○	○	○	○
15	○	○	○	○
16	○	○	○	○
17	○	○	○	○
18	○	○	○	○
19	○	○	○	○
20	○	○	○	○
21	○	○	○	○
22	○	○	○	○
23	○	○	○	○
24	○	○	○	○
25	○	○	○	○
26	○	○	○	○
27	○	○	○	○
28	○	○	○	○
29	○	○	○	○
30	○	○	○	○

MARKS | DO NOT WRITE IN THIS MARGIN

SECTION 2 — 70 marks

Attempt ALL questions

1. Will-o'-the-wisp is a ghostly flickering flame that appears over marshes. According to folklore, it lured travellers from well-trodden paths into treacherous marshes. It is now known to be caused by the spontaneous combustion of a mixture of methane and phosphorus(III) hydride formed from rotting vegetation in the depths of marshes.

   (a) Draw the shape of the phosphorus(III) hydride molecule.

   1

   (b) Phosphorus(III) hydride contains the element phosphorus and the electronic configuration of a phosphorus atom in its ground state can be represented as:

   1s   2s   2p   3s   3p

   (i) State the values of the four quantum numbers (*n*, *l*, *m* and *s*) for one of the electrons in the 3s orbital.

   2

   (ii) In what respect is the 1s orbital similar to and how does it differ from the 2s orbital?

   1

MARKS

1. **(b) (continued)**

   (iii) Explain how the above representation of the phosphorus atom conforms to Hund's rule.

   1

   (iv) In an excited state of a phosphorus atom, one of its electrons has the quantum numbers $n = 4$ and $l = 2$.

   In what type of orbital is this electron located?

   1

   (c) The concentration of methane gas in a 2 kg sample of air was found to be 5000 ppm.

   Calculate the number of moles of methane present.

   2

2. The nickel content of a compound can be determined by the three methods listed below.

   • Volumetric

   • Colorimetric

   • Gravimetric

   (a) In the volumetric method, a complexometric titration can be carried out.

   1·33g of hydrated nickel(II)sulfate, $NiSO_4 \cdot 6H_2O$, was weighed out accurately and made up to 100·0cm³ in a standard flask using distilled water.

   25·0cm³ portions were titrated against 0·110mol l⁻¹ EDTA solution using murexide indicator.

   The results of the complexometric titrations are shown in the table below.

	Rough titration	1st titration	2nd titration
Initial burette reading/cm³	0·0	12·0	23·7
Final burette reading/cm³	12·0	23·7	35·5
Volume of EDTA added/cm³	12·0	11·7	11·8

   (i) From the experimental results, calculate the percentage of nickel, by mass, in the hydrated nickel(II) sulfate sample.   3

MARKS | DO NOT WRITE IN THIS MARGIN

2. (a) (continued)

(ii) At the end point, the colour of the solution changes from yellow to violet.

Sketch the absorption curve for the violet solution below.   **1**

(b) A solution of nickel(II) sulfate is green in colour and so a pure sample of it can be used in a colorimetric method to determine the percentage of nickel in a solution whose concentration is not accurately known.

In the gravimetric method, a solution of nickel(II) sulfate reacts with dimethylglyoxime to produce a red solid. This can be filtered off, washed, dried and weighed to constant mass in order to determine the percentage of nickel by mass.

Considering all three methods involved, state one major factor in each method which could contribute to uncertainty in the final result.   **3**

(c) A nickel solution was passed through a flame and the light emitted observed through a spectroscope. A line in the emission spectrum for nickel was observed at a wavelength of 508 nm.

Calculate the energy, in kJ mol^{-1}, associated with this wavelength.   **2**

MARKS | DO NOT WRITE IN THIS MARGIN

3. The huge variety of colour we observe on planet Earth can be traced to the movement of electrons in an atom or molecule.

   Using your knowledge of chemistry, comment on how the colour we see is created by the movement of electrons.

   3

MARKS

4. Consider the following reaction.

$$CS_2(g) + 4H_2(g) \rightleftharpoons CH_4(g) + 2H_2S(g)$$

At 900 °C the equilibrium concentrations are:

$[CS_2] = 0.012 \, mol \, l^{-1}$ $[H_2] = 0.0020 \, mol \, l^{-1}$

$[H_2S] = 0.00010 \, mol \, l^{-1}$ $[CH_4] = 0.0054 \, mol \, l^{-1}$

(a) Write down the expression for the equilibrium constant, K, for this reaction.

1

(b) Calculate the value of the equilibrium constant, K, at 900 °C.

1

MARKS | DO NOT WRITE IN THIS MARGIN

5. Carbonated soft drinks contain carbon dioxide gas which dissolves in water to produce carbonic acid:

$$CO_2(aq) + H_2O(l) \rightleftharpoons H^+(aq) + HCO_3^-(aq) \qquad pK_a = 6{\cdot}35$$

(a) Identify the conjugate base in this reaction.   **1**

(b) Calculate the pH of a $0{\cdot}01\,mol\,l^{-1}$ solution of carbonic acid.   **2**

(c) A solution of carbonic acid was titrated with potassium hydroxide solution.

    (i) Suggest a suitable indicator to determine the end point of this reaction.   **1**

    (ii) The solution formed at the end of the titration was added to carbonic acid.

        Explain why the resultant solution can resist changes in pH even when small quantities of acid or base are added.   **2**

6. Burning magnesium continues to burn when placed in a gas jar of carbon dioxide according to the equation

$$2Mg(s) + CO_2(g) \rightarrow 2MgO(s) + C(s)$$

Substance	$S^o / J\,K^{-1}\,mol^{-1}$
$Mg(s)$	33·0
$CO_2(g)$	214·0
$MgO(s)$	27·0
$C(s)$	5·70

(a) Using the values from the table above, calculate ΔS^o for the reaction.   **1**

(b) Using the information below and your answer to (a), calculate ΔG^o for the burning of magnesium in carbon dioxide at 298 K.

$Mg(s) + \frac{1}{2}O_2(g) \rightarrow MgO(s)$ $\Delta H^o = -493\,kJ\,mol^{-1}$

$C(s) + O_2(g) \rightarrow CO_2(g)$ $\Delta H^o = -394\,kJ\,mol^{-1}$ **3**

MARKS | DO NOT WRITE IN THIS MARGIN

7. Consider the three reactions and their rate equations.

Reaction 1	$2N_2O_5 \rightarrow 4NO_2 + O_2$	Rate = $k[N_2O_5]$
Reaction 2	$2NO + Cl_2 \rightarrow 2NOCl$	Rate = $k[NO]^2[Cl_2]$
Reaction 3	$2NH_3 \rightarrow N_2 + 3H_2$	Rate = $k[NH_3]^0$

(a) What is the overall order of Reaction 2?   1

(b) The graph below was plotted using experimental results from one of the reactions.

Explain which of the reactions would give this graph.   1

(c) For Reaction 2, when the concentrations of NO and Cl_2 are both $0\cdot250\,mol\,l^{-1}$, the initial reaction rate is $1\cdot43 \times 10^{-6}\,mol\,l^{-1}\,s^{-1}$.

Use this information to calculate the rate constant, k, including the appropriate units.   2

MARKS | DO NOT WRITE IN THIS MARGIN

8. The following procedure was used to determine the mass of aspirin present in an aspirin tablet.

Step 1: Crush an aspirin tablet and heat with excess sodium hydroxide solution.

Aspirin
(1 mole = 180 g)

Step 2: Determine the excess sodium hydroxide by back titration using a standard solution of sulfuric acid.

$$2NaOH + H_2SO_4 \longrightarrow Na_2SO_4 + 2H_2O$$

An aspirin tablet was crushed and added to $50 \cdot 0\,cm^3$ of $1 \cdot 00\,mol\,l^{-1}$ sodium hydroxide solution. The solution was heated for 30 minutes. The cooled reaction mixture was then added to a $500\,cm^3$ standard flask, which was made up to the mark using deionised water. $25 \cdot 0\,cm^3$ samples were then titrated with $0 \cdot 10\,mol\,l^{-1}$ sulfuric acid until concordant results were obtained. The average titre was $12 \cdot 1\,cm^3$.

(a) Calculate the mass of aspirin in the tablet. **4**

(b) Suggest why a back titration technique has to be used to determine the mass of aspirin. **1**

8. **(continued)**

(c) In a separate experiment, a student synthesised a sample of aspirin by reacting 2-hydroxybenzoic acid with excess ethanoic anhydride. They reported a 74% yield.

2-hydroxybenzoic acid (1 mole = 138 g)   ethanoic anhydride   aspirin (1 mole = 180 g)

(i) Give a reason why the percentage yield is not 100%.   1

(ii) Calculate the mass of 2-hydroxybenzoic acid that should be reacted with excess ethanoic anhydride in order to produce 8 g of aspirin, assuming a 74% yield.   2

(iii) The sample of aspirin obtained was analysed by thin layer chromatography and melting point analysis. Explain how both techniques can be used to help determine the purity of the aspirin.   2

MARKS | DO NOT WRITE IN THIS MARGIN

8. **(continued)**

(d) Colorimetry can be used to determine the quantity of aspirin in a tablet. This is achieved by hydrolysing the aspirin to produce 2-hydroxybenzoic acid, which forms a coloured compound when reacted with iron(III) ions.

A method for the colorimetric determination of aspirin is shown below.

Colorimetric determination of aspirin

1. Weigh accurately about $0.4\,g$ of aspirin into a $100\,cm^3$ conical flask, add $10\,cm^3$ of $1.0\,mol\,l^{-1}$ sodium hydroxide solution and warm the mixture gently at $50\,°C$ for ten minutes.

2. Cool the solution and transfer quantitatively to a $500\,cm^3$ volumetric flask. Make up to the mark with deionised water. This is the stock solution which has a concentration equivalent to $0.8\,g\,l^{-1}$ of aspirin.

3. Use a burette to measure $10\,cm^3$ of stock solution into a $100\,cm^3$ volumetric flask. Make it up to volume with $0.02\,mol\,l^{-1}$ iron(III) chloride solution. This is standard solution A which has an equivalent aspirin concentration of $0.08\,g\,l^{-1}$. In a similar way make up further solutions and measure their absorbance using an appropriate filter on the colorimeter.

(i) Explain the meaning of the term "weigh accurately about $0.4\,g$".   1

(ii) Show by calculation that the stock solution has a concentration equivalent to $0.8\,g\,l^{-1}$.   1

(iii) Explain how the student would decide on "an appropriate filter" for this experiment.   1

(iv) Suggest an improvement that could be made to step 3 to improve the accuracy of the results.   1

9. Cyclohexanol can be converted into cyclohexene or cyclohexanone using different reagents as outlined below.

(a) Suggest a dehydrating agent that could be used to convert cyclohexanol into cyclohexene in reaction 1.   **1**

(b) **Using your knowledge of chemistry**, comment on how it could be established that the product of reaction 2 is cyclohexanone.   **3**

MARKS | DO NOT WRITE IN THIS MARGIN

10. Consider the following reaction sequence.

(a) HCN provides the CN⁻ ion which attacks a positive carbon atom in reagent **A**.

Which term best describes reagent **A**?   **1**

(b) Lactic acid in the form of lactate ions is dehydrogenated in the liver by the enzyme lactate dehydrogenase.

The diagram shows how one of the optical isomers of the lactate ion binds to an active site of lactate dehydrogenase.

art of the lactate dehydrogenase molecule

H
|
C ⋯⋯ OH
H₃C
CO_2^-

□ hydrophobic region
▲ hydrogen bonding region
● ionic bonding region

active site

(i) Draw a structure for the other optical isomer of the lactate ion.   **1**

(ii) Explain why this other optical isomer of the lactate ion cannot bind as efficiently to the active site of lactate dehydrogenase.   **1**

MARKS | DO NOT WRITE IN THIS MARGIN

11. A mixture of butan-1-ol and butan-2-ol can be synthesised from 1-bromobutane in a two-stage process.

$$CH_3CH_2CH_2CH_2Br \xrightarrow[\text{Stage 1}]{KOH/C_2H_5OH} CH_3CH_2CH=CH_2 \xrightarrow[\text{Stage 2}]{H_2O/H^+} \begin{array}{c} \text{butan-1-ol} \\ + \\ \text{butan-2-ol} \end{array}$$

(a) But-1-ene is produced after Stage 1. State the name of the other product formed at Stage 1.

1

(b) The bonding in but-1-ene can be described in terms of sp^2 and sp^3 hybridisation and sigma and pi bonds.

(i) What is meant by sp^2 hybridisation?

1

(ii) What is the difference in the way atomic orbitals overlap to form sigma and pi bonds?

1

(c) Draw a structural formula for the major product of Stage 2.

1

(d) 1-Bromobutane reacts with hydroxide ions to form butan-1-ol by an S_N2 mechanism.

Using structural formulae, draw the mechanism for S_N2 reaction of 1-bromobutane with hydroxide ions.

2

MARKS | DO NOT WRITE IN THIS MARGIN

12. A compound **X** containing only carbon, hydrogen and oxygen was subjected to elemental analysis. Complete combustion of 1·76 g of **X** gave 3·52 g of carbon dioxide and 1·44 g of water. No other product was formed.

(a) (i) Calculate the masses of carbon and hydrogen in the original sample and hence deduce the mass of oxygen present.   **2**

(ii) Show, by calculation, that the empirical formula of compound **X** is C_2H_4O.   **1**

(b) Given that the relative molecular mass of compound **X** is 88, deduce its molecular formula.   **1**

MARKS | DO NOT WRITE IN THIS MARGIN

12.   (continued)

(c)   A low-resolution proton NMR spectrum of compound **X** is shown.

Analysis of this spectrum produced the data shown in the table below.

Chemical shift	Relative area under the peak
0·9	3
1·6	2
2·6	2
0·9 1·6 2·6 10·8	1

(i)   Using information from the table and your answer to (b), draw a structural formula for compound **X** and give its systemic name.   **2**

(ii)   What is the function of 'TMS' (tetramethylsilane) in the proton NMR spectrum?   **1**

[END OF MODEL PAPER]

ADDITIONAL SPACE FOR ANSWERS AND ROUGH WORK

ADDITIONAL SPACE FOR ANSWERS AND ROUGH WORK

ADVANCED HIGHER FOR CfE

Model Paper 3

Whilst this Model Paper has been specially commissioned by Hodder Gibson for use as practice for the Advanced Higher (for Curriculum for Excellence) exams, the key reference document remains the SQA Specimen Paper 2015.

HODDER
GIBSON
LEARN MORE

National
Qualifications
MODEL PAPER 3

Chemistry
Section 1 — Questions

Duration — 2 hours 30 minutes

Instructions for the completion of Section 1 are given on *Page two* of your question and answer booklet.

Record your answers on the answer grid on *Page three* of your question and answer booklet.

Reference may be made to the Chemistry Higher and Advanced Higher Data Booklet.

Before leaving the examination room you must give your question and answer booklet to the Invigilator; if you do not you may lose all the marks for this paper.

SECTION 1 — 30 marks

Attempt ALL questions

1. The electronic configuration of $Cu^+(g)$ is

   A $4s^1 3d^9$

   B $4s^2 3d^8$

   C $4s^0 3d^{10}$

   D $4s^1 3d^9$.

The following two questions refer to the absorption spectrum of $[Ti(H_2O)_6]^{3+}$ below.

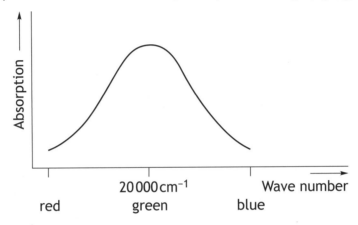

2. The colour which would be observed is

   A   blue

   B   green

   C   purple

   D   yellow.

3. The colour is caused by

   A   electrons absorbing energy to jump from one d orbital to another d orbital

   B   excited electrons dropping to the ground state

   C   electrons absorbing energy to jump from the first to the second shell

   D   electrons absorbing energy to jump from the second to the third shell.

4. The correct formula for the tetraamminedichlorocopper(II) complex is

   A $[Cu(NH_3)_4Cl_2]^{2-}$

   B $[Cu(NH_3)_4Cl_2]$

   C $[Cu(NH_3)_4Cl_2]^{2+}$

   D $[Cu(NH_3)_4Cl_2]^{4+}$.

Page two

5. When one mole of phosphorus pentachloride was heated to 523 K in a closed vessel, 50% dissociated as shown

$$PCl_5(g) \rightleftharpoons PCl_3(g) + Cl_2(g)$$

How many moles of gas were present in the equilibrium mixture?

A   0·5

B   1·0

C   1·5

D   2·0

6. Which of the following would **not** be suitable to act as a buffer solution?

A   Boric acid and sodium borate

B   Nitric acid and sodium nitrate

C   Benzoic acid and sodium benzoate

D   Propanoic acid and sodium propanoate

7. Which of the following $0.01 \, mol \, l^{-1}$ aqueous solutions has the highest pH value?

A   Sodium fluoride

B   Sodium benzoate

C   Sodium propanoate

D   Sodium methanoate

8. An acid is a substance which

A   donates a proton leaving a conjugate acid

B   donates a proton leaving a conjugate base

C   accepts a proton leaving a conjugate acid

D   accepts a proton leaving a conjugate base.

9. The pH ranges over which some indicators change colour are shown below.

Which line in the table shows the indicator most suitable for the titration of hydrochloric acid with ammonia solution?

	Indicator	pH range
A	Methyl orange	4·2–6·3
B	Bromothymol blue	6·0–7·6
C	Phenol red	6·8–8·4
D	Phenolphthalein	8·3–10·0

10. Which of the following reactions would show the greatest decrease in entropy?

A $H_2(g) + F_2(g) \rightarrow 2HF(g)$

B $KNO_3(s) \rightarrow KNO_2(s) + \frac{1}{2}O_2(g)$

C $CO_3^{2-}(aq) + 2H^+(aq) \rightarrow H_2O(l) + CO_2(g)$

D $CO_3^{2-}(aq) + CO_2(g) + H_2O(l) \rightarrow 2HCO_3^-(aq)$

11. Which line in the table is correct for the enthalpy change and entropy change when steam condenses?

	ΔH	ΔS
A	+ve	+ve
B	+ve	–ve
C	–ve	–ve
D	–ve	+ve

12. Which of the following graphs shows the variation in $\Delta G°$ with temperature for a reaction which is always feasible?

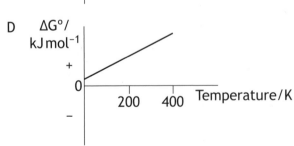

13. For the reaction

$2NO(g) + Cl_2(g) \rightarrow 2NOCl(g)$

the rate equation is

rate = $k[NO][Cl_2]$

The overall order of this reaction is

A   1

B   2

C   3

D   5.

14. What volume of **water** needs to be added to $25\,cm^3$ of $0\cdot2\,mol\,l^{-1}$ calcium chloride solution to produce a solution with a **chloride** ion concentration of $0\cdot1\,mol\,l^{-1}$?

A $25\,cm^3$

B $50\,cm^3$

C $75\,cm^3$

D $100\,cm^3$

15. Which of the following best describes the bonding in ethane?

A sp^2 hybridisation of the carbon atoms giving sigma bonds only

B sp^2 hybridisation of the carbon atoms giving sigma and pi bonds

C sp^3 hybridisation of the carbon atoms giving sigma bonds only

D sp^3 hybridisation of the carbon atoms giving sigma and pi bonds

16. $KI_{(aq)} + Pb(NO_3)_{2\,(aq)} \rightarrow PbI_{2\,(s)} + KNO_{3\,(aq)}$

Which of the following would be the most appropriate technique for obtaining a dry sample of PbI_2?

A   evaporation of the product mixture

B   distillation of the product mixture

C   vacuum filtration of the product mixture

D   refluxing the product mixture

17. In the homologous series of alcohols, increase in chain length from CH_3OH to $C_{10}H_{21}OH$ is accompanied by

    A   increased volatility and increased solubility in water

    B   increased volatility and decreased solubility in water

    C   decreased volatility and decreased solubility in water

    D   decreased volatility and increased solubility in water.

18. Which of the following is **not** caused by hydrogen bonding?

    A   The low density of ice compared to water

    B   The solubility of methoxymethane in water

    C   The higher boiling point of methanol compared to ethane

    D   The higher melting point of hydrogen compared to helium

19. Melting point determination and thin layer chromatography are useful techniques for assessing the purity of a compound. Which line in the table below shows the results expected from a pure compound, which is known to have a melting point of 128°C?

	Melting point range (°C)	Number of spots shown on a chromatography plate
A	92·1–101·7	3
B	120·2–128·0	1
C	127·5–127·9	1
D	127·8–130·1	2

20. Which of the following compounds would liberate one mole of hydrogen gas if one mole of it reacts with two moles of sodium?

    A C_2H_5OH

    B $HOCH_2CH_2OH$

    C CH_3COOH

    D CH_3CHO

21. Mandelic acid has two optical isomers, **X** and **Y**. The table shows the rotation of plane polarised light caused by various solutions of **X** and **Y**.

Volume of 0·1 mol l^{-1} X/cm^3	Volume of 0·1 mol l^{-1} Y/cm^3	Volume of water/cm^3	Observed rotation/°
100	0	0	+158
50	0	50	+79
50	50	0	0
0	100	0	−158

What would be the observed rotation for a solution containing 25 cm^3 of 0·1 mol l^{-1} **X** and 75 cm^3 of 0·1 mol l^{-1} **Y**?

A   −79°

B   −39·5°

C   +39·5°

D   79°

22. Carbonyl groups in aldehydes and ketones react with HCN and the product can then be hydrolysed forming a 2-hydroxycarboxylic acid as shown in the equation below.

methanal   2-hydroxyethanoic acid

When the final product is 2-hydroxy-2-methylbutanoic acid, the starting carbonyl compound is

A   propanal

B   propanone

C   butanal

D   butanone.

23.

is the skeletal formula for

A   3,4,4-trimethylbutan-1-ol

B   2,3-dimethylpentan-5-ol

C   3,4-dimethylpentan-1-ol

D   4,5-dimethylhexan-1-ol.

24. One mole of which of the following compounds will react with the largest volume of $1\,mol\,l^{-1}$ hydrochloric acid?

    A CH_3NHCH_3

    B $H_2NCH_2NH_2$

    C $HOOCCH_2NH_2$

    D

    HOOC—⬡—NH_2

25. Which of the following is **not** an example of a hydrolysis reaction?

    A $C_2H_4 + H_2O \rightarrow C_2H_5OH$

    B $CH_3CN + 2H_2O \rightarrow CH_3COOH + NH_3$

    C $CH_3COOCH_3 + H_2O \rightarrow CH_3COOH + CH_3OH$

    D $C_6H_5COOCH_3 + H_2O \rightarrow C_6H_5COOH + CH_3OH$

26. Spectral studies of an organic compound indicated the presence of a di-substituted benzene ring, two methyl groups and a molecular weight of 134.

    Which of the following is a possible structure for the compound?

    A CH_3

    $COCH_3$

    B CH_3

    CHO

    CH_3

    C CH_3CH_2

    $COCH_3$

    D CH_3

    CH_2CHO

27. Which of the following compounds has a geometric isomer?

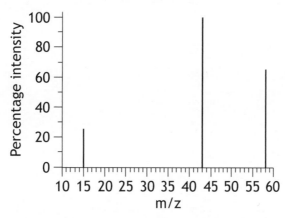

28. A simplified mass spectrum of an organic compound is shown.

Which of the following compounds produces this spectrum?

A   Propane

B   Propan-1-ol

C   Propan-2-ol

D   Propanone

29. An analysis of an organic compound found in meteorite rocks shows the following percentage composition by mass:

C = 37·5%   H = 12·5%   O = 50%

The empirical (simplest) formula for the compound is

A CH_4O

B C_3HO_4

C $C_3H_{12}O_3$

D CH_2O_2.

30. Salbutamol is used to treat asthma. It behaves like the body's natural active compound by binding to receptors on the muscles of the air passages. This relaxes the muscles and gives relief from breathing difficulties. Salbutamol is

A   an agonist

B   an antagonist

C   a pharmacaphore

D   a receptor.

[END OF SECTION 1. NOW ATTEMPT THE QUESTIONS IN SECTION 2
OF YOUR QUESTION AND ANSWER BOOKLET.]

FOR OFFICIAL USE

National
Qualifications
MODEL PAPER 3

Mark

Chemistry
Section 1 — Answer Grid
and Section 2

Duration — 2 hours 30 minutes

Fill in these boxes and read what is printed below.

Full name of centre

Town

Forename(s)

Surname

Number of seat

Date of birth

Day   Month   Year

D D   M M   Y Y

Scottish candidate number

Reference may be made to the Chemistry Higher and Advanced Higher Data Booklet.

Total marks — 100

SECTION 1 — 30 marks

Attempt ALL questions.

Instructions for completion of Section 1 are given on *Page two*.

SECTION 2 — 70 marks

Attempt ALL questions

Write your answers clearly in the spaces provided in this booklet. Additional space for answers and rough work is provided at the end of this booklet. If you use this space you must clearly identify the question number you are attempting. Any rough work must be written in this booklet. You should score through your rough work when you have written your final copy.

Use **blue** or **black** ink.

Before leaving the examination room you must give this booklet to the Invigilator; if you do not, you may lose all the marks for this paper.

SECTION 1— 30 marks

The questions for Section 1 are contained in the question paper on *Page four*.
Read these and record your answers on the answer grid on *Page three* opposite.
Do NOT use gel pens.

1. The answer to each question is **either** A, B, C or D. Decide what your answer is, then fill in the appropriate bubble (see sample question below).

2. There is **only one correct** answer to each question.

3. Any rough working should be done on the additional space for answers and rough work at the end of this booklet.

Sample Question

To show that the ink in a ball-pen consists of a mixture of dyes, the method of separation would be:

    A   fractional distillation

    B   chromatography

    C   fractional crystallisation

    D   filtration.

The correct answer is **B**—chromatography. The answer **B** bubble has been clearly filled in (see below).

Changing an answer

If you decide to change your answer, cancel your first answer by putting a cross through it (see below) and fill in the answer you want. The answer below has been changed to **D**.

If you then decide to change back to an answer you have already scored out, put a tick (✓) to the **right** of the answer you want, as shown below:

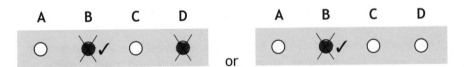

SECTION 1 — Answer Grid

	A	B	C	D
1	○	○	○	○
2	○	○	○	○
3	○	○	○	○
4	○	○	○	○
5	○	○	○	○
6	○	○	○	○
7	○	○	○	○
8	○	○	○	○
9	○	○	○	○
10	○	○	○	○
11	○	○	○	○
12	○	○	○	○
13	○	○	○	○
14	○	○	○	○
15	○	○	○	○
16	○	○	○	○
17	○	○	○	○
18	○	○	○	○
19	○	○	○	○
20	○	○	○	○
21	○	○	○	○
22	○	○	○	○
23	○	○	○	○
24	○	○	○	○
25	○	○	○	○
26	○	○	○	○
27	○	○	○	○
28	○	○	○	○
29	○	○	○	○
30	○	○	○	○

MARKS | DO NOT WRITE IN THIS MARGIN

SECTION 2 — 70 marks

Attempt ALL questions

1. Atomic spectroscopy is a useful analytical tool for identifying and quantifying the elements present in a sample. It also provides information about atomic structure.

   (a) When a high voltage is applied to a lamp filled with helium gas, a line of red light, wavelength 706 nm, is observed through a spectroscope.

      (i) Explain how the line of red light is produced. **2**

      (ii) Calculate the energy, in kJ mol^{-1}, associated with this wavelength. **2**

   (b) A helium atom has two electrons in its ground state. One of the electrons can be described by the four quantum numbers 1, 0, 0, $+\frac{1}{2}$. What four quantum numbers describe the other electron? **1**

MARKS | DO NOT WRITE IN THIS MARGIN

1. **(continued)**

(c)   (i)   Using orbital box notation, write the electronic configuration for a phosphorus atom in its ground state.   1

(ii)   Explain how your answer is consistent with Hund's rule.   1

(iii)   When a phosphorus atom becomes excited an electron can move to the 4s orbital.

What four quantum numbers describe the excited electron?   1

(d)   Mercury atoms are much larger than those of helium or phosphorus. A small section of the atomic emission spectrum for mercury is shown below.

400 nm   700 nm

Why does this spectrum have multiple lines?   1

(e)   Sir Isaac Newton was thought to have suffered from mercury poisoning and when his hair was analysed the level of mercury was found to be 73 ppm. The unit ppm stands for parts per million and refers to 1 mg per kg.

Calculate the number of moles of mercury in a 1.0×10^{-6} g sample of Newton's hair.   2

MARKS | DO NOT WRITE IN THIS MARGIN

2. A common dietary supplement taken by athletes and slimmers is called chromium picolinate $[Cr(pic)_3]$. The structure of the picolinate ion, pic, is

(a) What feature of the picolinate ion makes it suitable for use as a ligand?   **1**

(b) In the body, it is thought that the chromium in $[Cr(pic)_3]$ is changed into chromium(VI) by the action of hydrogen peroxide.

(i) What is the oxidation state of chromium in $[Cr(pic)_3]$?   **1**

(ii) What is the role of hydrogen peroxide in this reaction?   **1**

(c) A simpler complex of chromium is $[Cr(CN)_6]^{4-}$. What is its systematic name?   **1**

MARKS | DO NOT WRITE IN THIS MARGIN

3. The dicarboxylic acid, oxalic acid, has molecular formula $H_2C_2O_4$.

It can be prepared by reacting calcium oxalate with sulfuric acid.

$$H_2SO_4(aq) + CaC_2O_4(s) + xH_2O(l) \rightarrow CaSO_4 \cdot xH_2O(s) + H_2C_2O_4(aq)$$

(a) Draw a structural formula for oxalic acid.   **1**

(b) 4·94 g of $CaSO_4 \cdot xH_2O$ was dehydrated to produce 3·89 g of $CaSO_4$.

Determine the value of x.   **2**

(c) The equation for the reaction between oxalic acid and sodium hydroxide solution is

$$H_2C_2O_4(aq) + 2NaOH(aq) \rightarrow Na_2C_2O_4(aq) + 2H_2O(l)$$

A student used a standard solution of 0·0563 mol l^{-1} oxalic acid to standardise 20·0 cm^3 of approximately 0·1 mol l^{-1} sodium hydroxide solution.

The raw results for the titration are given in the table.

	1st attempt	2nd attempt	3rd attempt
Final burette reading/cm^3	17·2	33·8	16·6
Initial burette reading/cm^3	0·0	17·2	0·1
Titre/cm^3	17·2	16·6	16·5

Calculate the accurate concentration of the sodium hydroxide solution.   **2**

(d) Oxalic acid is a primary standard but sodium hydroxide is not. State **one** characteristic of sodium hydroxide that makes it unsuitable as a primary standard.   **1**

MARKS

4. Before a pH electrode can be used to measure the pH of a solution it must be calibrated by placing it in a series of buffer solutions of known pH.

One buffer solution used to calibrate a pH electrode was made by dissolving 2·24 g of potassium propanoate, C_2H_5COOK, in $250\,cm^3$ of $0·20\,mol\,l^{-1}$ propanoic acid.

(a) What property of buffer solutions makes them ideal for calibrating pH electrodes?

1

(b) Calculate the pH of the buffer solution described above.

3

5. The following table of results was obtained for the reaction below.

$$H_2O_2(aq) + 2HI(aq) \rightarrow 2H_2O(l) + I_2(aq)$$

Experiment	$[H_2O_2]/mol\,l^{-1}$	$[HI]/mol\,l^{-1}$	Initial rate$/mol\,l^{-1}s^{-1}$
1	$3 \cdot 2 \times 10^{-4}$	$4 \cdot 1 \times 10^{-4}$	$4 \cdot 3 \times 10^{-9}$
2	$6 \cdot 4 \times 10^{-4}$	$4 \cdot 1 \times 10^{-4}$	$8 \cdot 6 \times 10^{-9}$
3	$3 \cdot 2 \times 10^{-4}$	$8 \cdot 2 \times 10^{-4}$	$8 \cdot 6 \times 10^{-9}$
4	$6 \cdot 4 \times 10^{-4}$	$8 \cdot 2 \times 10^{-4}$	$1 \cdot 72 \times 10^{-8}$

(a) The order of reaction with respect to H_2O_2 is 1st order. Determine the order of reaction with respect to HI.

1

(b) Write the rate equation for the reaction.

1

(c) Calculate a value for the rate constant, k, including the appropriate units.

2

MARKS | DO NOT WRITE IN THIS MARGIN

6. Molten iron, made in a blast furnace, often contains sulfur and phosphorus impurities which must be removed.

   Bubbling carbon dioxide gas through molten iron removes the sulfur.

   The carbon dioxide gas is produced by the decomposition of calcium carbonate.

$$CaCO_3(s) \rightarrow CaO(s) + CO_2(g)$$

Substance	Standard enthalpy of formation, $\Delta H^\circ / kJ\,mol^{-1}$	Standard entropy, $S^\circ / J\,K^{-1}\,mol^{-1}$
CO_2	−393·5	213·8
$CaCO_3$	−1206·9	92·9
CaO	−635·1	38·1

For the decomposition of calcium carbonate, use the data in the table to calculate

(a) the standard enthalpy change, ΔH°, in $kJ\,mol^{-1}$ 1

(b) the standard entropy change, ΔS°, in $J\,K^{-1}\,mol^{-1}$ 1

(c) the theoretical temperature at which the reaction just becomes feasible.   3

7. The commercial medicine, paracetamol, can be created as shown in the reaction sequence below.

Step 1: Nitration

Step 2: Reduction

74% yield

Step 3: Amide formation

Paracetamol

(a) State the name of an alternative reducing agent to $NaBH_4$ that could be used to carry out Step 2.   **1**

(b) State the chemical formula for paracetamol.   **1**

MARKS | DO NOT WRITE IN THIS MARGIN

7. (continued)

(c) A student carried out Step 3 by reacting 4-aminophenol with excess ethanoic anhydride. The paracetamol obtained was purified by recrystallisation. A small sample was then used to determine the melting point whilst another sample was used for thin layer chromatography.

(i) Assuming a 60% yield for Step 3, calculate the minimum mass of 4-aminophenol required to react with ethanoic anhydride to produce 1·10 g of paracetamol.

3

(ii) Describe the steps the student would take to recrystallise the paracetamol.

2

(d) The quantity of paracetamol in a sample can be determined by refluxing an accurate mass of paracetamol with sulfuric acid to form 4-aminophenol, followed by titration with the oxidising agent ammonium cerium(IV) sulfate.

4-aminophenol   Ethanoic acid

Draw a diagram of the apparatus used to carry out the reflux as described.

2

MARKS | DO NOT WRITE IN THIS MARGIN

8. The percentage of iron in an iron supplement can be determined by colorimetry. An extract from a laboratory manual for preparing a sample for colorimetric analysis is shown below.

Iron tablet preparation

1. Place 1 iron tablet into a beaker along with $25\,cm^3$ of $6\,mol\,l^{-1}$ HCl (aq) and boil gently for 15 min.

2. Filter into a $100\,cm^3$ volumetric flask.

3. Add $3\,cm^3$ o-phenanthroline, and dilute to $100\,cm^3$ with water.

4. Measure the absorbance of this solution at 508 nm.

The iron in the tablet reacts with o-phenanthroline as shown:

o-phenanthroline λ_{max} = 508 nm

(a) State the purpose of the compound, o-phenanthroline.   1

(b) Describe all further steps a student would have to take in order to determine the concentration of iron in the iron tablet using the technique of colorimetry.   3

MARKS | DO NOT WRITE IN THIS MARGIN

8. (continued)

(c) The recommended daily allowance (RDA) of iron is 14·8 mg.

An iron tablet was treated as per steps 1–4 in the guide above. The 100 cm^3 solution prepared was found to have an iron concentration of 4·67 x 10^{-3} mol l^{-1}. Calculate the % of RDA supplied by this iron tablet.   **2**

(d) An alternative method of iron determination involves titrating with a standard solution of potassium permanganate, since permanganate ions react with iron (II) according to the following equation:

$$MnO_4^-{}_{(aq)} + 8H^+{}_{(aq)} + 5Fe^{2+}{}_{(aq)} \rightarrow Mn^{2+}{}_{(aq)} + 5Fe^{3+}{}_{(aq)} + 4H_2O_{(l)}$$

Another brand of iron tablet was treated as per the instructions above. 10 cm^3 aliquots from the 100 cm^3 iron tablet solution were titrated with 0·002 mol l^{-1} standard potassium permanganate solution. The average titre was found to be 20·3 cm^3. Calculate the mass of iron present in this tablet.   **3**

MARKS | DO NOT WRITE IN THIS MARGIN

9. In the winter of 1987 over a hundred people became extremely ill within hours after dining on cultured blue mussels in restaurants around Prince Edward Island in Canada. After much research, chemists eventually isolated and identified the toxin to be domoic acid, a structural formula of which is shown below.

Using your knowledge of chemistry, suggest how domoic acid could be isolated and identified from contaminated mussels.

3

MARKS

10. 1-Bromobutane and 2-bromo-2-methylpropane are isomers which can be converted to their corresponding alcohols using hydroxide ions.

(a) The conversion of 2-bromo-2-methylpropane to 2-methylpropan-2-ol involves an S_N1 mechanism.

Outline the two steps in this mechanism using structural formulae.   **2**

(b) Explain why the conversion of 1-bromobutane to butan-1-ol is unlikely to proceed by an S_N1 mechanism.   **1**

MARKS | DO NOT WRITE IN THIS MARGIN

11. Blood clotting is a natural chemical reaction which helps to stop bleeding. In some cases, the blood can clot when not required leading to blockages in the veins. A potential treatment to stop blood clotting like this involves blocking the enzyme known as FXa which is one enzyme responsible for creating blood clots.

The structure of a commercial FXa inhibitor, Rivaroxaban, is shown below.

(a) Explain whether this molecule is acting as an agonist or antagonist.   2

MARKS

11.   (continued)

(b) Rivaroxaban can be synthesised by the reaction of the amine shown below with a carboxylic acid.

(i)   State whether this is a primary, secondary or tertiary amine.

1

(ii) Explain why solutions of this compound would be expected to have an alkaline pH.

1

(iii) Draw a structural formula for the carboxylic acid that can react with this amine to form Rivaroxaban.

1

12. The following simplified spectra were obtained for a pure organic compound containing carbon, hydrogen and oxygen only.

The proton NMR spectrum was analysed and the data are presented in the table below.

Chemical shift	Area under the peak	H atom ratio
1·1	6·9	X
2·5	4·5	Y
9·8	2·3	Z

(a) Identify the functional group causing the absorption at 1730 cm^{-1} in the infrared spectrum.

1

MARKS | DO NOT WRITE IN THIS MARGIN

12. **(continued)**

(b) Identify a possible ion fragment responsible for the peak at m/z 29 in the mass spectrum.

1

(c) From the proton NMR spectrum what is the whole number ratio of hydrogen atoms, X : Y : Z?

1

(d) From the information above draw a structural formula for this compound.

1

MARKS

DO NOT WRITE IN THIS MARGIN

13. Hybridisation of atomic orbitals is necessary to explain the bonding in hydrocarbon families such as the alkanes, alkenes and aromatic hydrocarbons.

   Using your knowledge of chemistry, discuss this statement.

3

[END OF MODEL PAPER]

ADDITIONAL SPACE FOR ANSWERS AND ROUGH WORK

ADDITIONAL SPACE FOR ANSWERS AND ROUGH WORK

[BLANK PAGE]

DO NOT WRITE ON THIS PAGE

SQA AND HODDER GIBSON ADVANCED HIGHER FOR CfE CHEMISTRY 2015

SECTION 1

Question	Response	Mark
1.	D	1
2.	A	1
3.	C	1
4.	B	1
5.	C	1
6.	B	1
7.	D	1
8.	D	1
9.	D	1
10.	B	1
11.	C	1
12.	D	1
13.	A	1
14.	B	1
15.	C	1
16.	B	1
17.	D	1
18.	D	1
19.	B	1
20.	A	1
21.	D	1
22.	A	1
23.	C	1
24.	C	1
25.	A	1
26.	B	1
27.	C	1
28.	A	1
29.	D	1
30.	A	1

SECTION 2

Question			Expected response	Max mark	Additional guidance
1.	**(a)**		Lithium	1	
	(b)		Energy absorbed resulted in electrons within atoms being promoted to higher energy level. **(1 mark)** When the electron falls back to its original level, energy is emitted in the form of a photon whose wavelength corresponds to red light. **(1 mark)**	2	
	(c)		$E = \dfrac{Lhc}{(1000)\,\lambda}$ or $= \dfrac{6\cdot02 \times 10^{23} \times 6\cdot63 \times 10^{-34} \times 3 \times 10^{8}}{670\cdot7 \times 10^{-9} \times (1000)}$ **(1 mark)** $= 179\,kJ\,mol^{-1}$ **(1 mark)**	2	1000 may or may not be included at this step. Significant figure rule applies. Accept 178·53, 178·5 or 180.
2.	**(a)**		ClO^{-} (aq), or hypochlorite ion	1	
	(b)		$K_a = \dfrac{[H_3O^+][ClO^-]}{[HClO^-]}$	1	allow H^+
	(c)		$ClO^-(aq) + H^+(aq) \rightleftharpoons HClO(aq)$ or The H^+(aq) are removed by the conjugate base from the water equilibrium **(1 mark)** and $H_2O(\ell) \rightleftharpoons H^+(aq) + OH^-(aq)$ or This causes the water equilibrium to shift to the right-hand side producing excess OH^-(aq) and hence pH > 7. or Produces an excess of hydroxide ions. **(1 mark)**	2	State symbols not required.

Question			Expected response	Max mark	Additional guidance
3.			**The whole candidate response should first be read to establish its overall quality in terms of accuracy and relevance to the problem/situation presented.** There may be strengths and weaknesses in the candidate response: *assessors should focus as far as possible on the strengths, taking account of weaknesses (errors or omissions) only where they detract from the overall answer in a significant way, which should then be taken into account when determining whether the response demonstrates **reasonable, limited or no** understanding.* **Assessors should use their professional judgement to apply the guidance below to the wide range of possible candidate responses.**	3	This open-ended question requires comment on **why β–carotene is orange.** Candidate responses are expected to make comment on the basis of relevant chemistry ideas/concepts which might include one or more of: talk about conjugation, conjugated systems, chromophores, absorption of light, complementary colours, promotion of electrons, σ and σ^*, π and π^*, longer conjugation tends towards red, HOMO and LUMO.
			3 marks: The candidate has demonstrated a **good** conceptual understanding of the chemistry involved, providing a logically correct response to the problem/situation presented. This type of response might include a statement of principle(s) involved, a relationship or equation, and the application of these to respond to the problem/situation. This does not mean the answer has to be what might be termed an "excellent" answer or a "complete" one.		In response to this question, a **good** understanding might be demonstrated by a candidate response that: • makes comments based on one relevant chemistry idea/concept, in a **detailed/developed** response that is **correct or largely correct** (any weaknesses are minor and do not detract from the overall response) **or** • makes comments based on a range of relevant chemistry ideas/concepts, in a response that is **correct or largely correct** (any weaknesses are minor and do not detract from the overall response) **or** • otherwise demonstrates a good understanding of the chemistry involved.

Question			Expected response	Max mark	Additional guidance
			2 marks: The candidate has demonstrated a **reasonable** understanding of the chemistry involved, showing that the problem/situation is understood. This type of response might make some statement(s) that is/are relevant to the problem/situation, for example, a statement of relevant principle(s) or identification of a relevant relationship or equation.		In response to this question, a **reasonable** understanding might be demonstrated by a candidate response that: • makes comments based on one or more relevant chemistry idea(s)/concept(s), in a response that is **largely correct** but has **weaknesses** which detract to a small extent from the overall response **or** • otherwise demonstrates a reasonable understanding of the chemistry involved.
			1 mark: The candidate has demonstrated a **limited** understanding of the chemistry involved, showing that a little of the chemistry that is relevant to the problem/situation is understood. The candidate has made some statement(s) that is/are relevant to the problem/situation.		In response to this question, a **limited** understanding might be demonstrated by a candidate response that: • makes comments based on one or more relevant chemistry idea(s)/concept(s), in a response that has **weaknesses** which detract to a large extent from the overall response **or** • otherwise demonstrates a limited understanding of the chemistry involved.
			0 marks: The candidate has demonstrated **no** understanding of the chemistry that is relevant to the problem/situation. The candidate has made no statement(s) that is/are relevant to the problem/situation.		Where the candidate has only demonstrated knowledge and understanding of chemistry **that is not relevant to the problem/situation presented**, 0 marks should be awarded.
4.	(a)	(i)	+4, 4, 4+ , IV, four.	1	−4, 4− would not be accepted.
		(ii)	 1s 2s 2p 3s 3p 3d	1	Box labels not required. Noble gas start accepted in this case. Full or half headed arrows accepted, but not vertical lines. follow through
		(iii)	The three d orbitals are all filled singly (with parallel spins).	1	follow through
	(b)	(i)	Degenerate means that the orbitals are of equal energy.	1	

Question			Expected response	Max mark	Additional guidance
		(ii)	p-orbital	1	axes not necessary any orientation accepted
	(c)	(i)	Several solutions of accurate permanganate concentration are made up and the absorbance of each is measured. **(1 mark)** A calibration curve of concentration vs absorbance is drawn. **(1 mark)** The absorbance of the solution of unknown permanganate concentration is measured and the calibration graph is then used to determine the concentration corresponding to this absorbance. **(1 mark)**	3	

		(ii)	$[MnO_4^-] = 4 \cdot 25 \times 10^{-4}$ mol l^{-1}
			    moles $= 0 \cdot 25 \times 4 \cdot 25 \times 10^{-4}$
			        $= 1 \cdot 0625 \times 10^{-4}$ moles
			
			moles Mn $= 1 \cdot 0625 \times 10^{-4}$ moles
			mass of Mn $= 1 \cdot 0625 \times 10^{-4} \times 54 \cdot 9$ **(1 mark)**
			        $= 5 \cdot 833 \times 10^{-3}$ g
			    % Mn $= 1 \cdot 7\%$ **(1 mark)**
				2	1 mark for concept of $54 \cdot 9 \times$ calculated moles of manganese.
					
					1 mark for arithmetic.
					Significant figure rule applies.
					
					Accept 2, 1·67, 1·667.

5.	(a)		The mass must be in the region of 0·4 g and the exact mass must be known.	1	
	(b)		Pour mixture into standard flask, rinse beaker with distilled water and add rinsings to standard flask. **(1 mark)** and Repeat, mix, make to mark with distilled water. **(1 mark)**	2	
	(c)	(i)	low GFM or unstable in air or absorbs moisture or is not a primary standard	1	

Question			Expected response	Max mark	Additional guidance
		(ii)	Initial moles of HCl $0.02 \times 1 = 0.02$ mol Ave titre = 12.65 No. moles NaOH = $0.1 \times 0.01265 = 0.001265$ = no. moles HCl left unreacted in 10 cm³ $10 \times 0.001265 = 0.01265$ in 100 cm³ **(1 mark)** Moles of acid reacting $0.02 - 0.01265 = 0.00735$ mol **(1 mark)** Mass of $CaCO_3$ is $(0.00735/2) \times 100.1$ = 0.3679 g **(1 mark)** % $CaCO_3$ $(0.3679/0.390) \times 100$ = 94.3% **(1 mark)**	4	1 mark concept of the scaling × 10. 1 mark concept of subtraction to calculate moles of acid reacting. 1 mark concept for stoichiometry 2 : 1. 1 mark for arithmetic. Significant figure rule applies. 94, 94.33, 94.333.
6.	(a)		$C_{17}H_{19}N_3O_3S$	1	Can be given in any order.
	(b)		A racemate or racemic mixture.	1	
	(c)		Antagonist	1	
	(d)	(i)	In an acidic environment, such as the stomach, the inactive isomer would be converted into the active one. or Cheaper to have a racemic mixture as you don't have to separate enantiomers.	1	
		(ii)	The OH or hydroxyl group.	1	
7.	(a)		$\Delta S = \sum S°\text{products} - \sum S°\text{reactants}$ $= (152 + 2 \times 189) - (77 + 4 \times 174)$ **(1 mark)** $= (530) - (773)$ $= -243$ J K⁻¹ mol⁻¹ **(1 mark)**	2	
	(b)		$\Delta G° = \Delta H° - T\Delta S°$ $= (-244) - 298(-0.243)$ **(1 mark)** $= (-244) - (-72.414)$ $= -171.6$ (kJ mol⁻¹) **(1 mark)** Yes the reaction is feasible. **(1 mark)**	3	Working must be shown. Follow through applies. Units need not be given. Answer in J acceptable.
8.	(a)	(i)	Solvent extraction.	1	
		(ii)	Drain layers into two separate beakers **(1 mark)** Return lower/aqueous layer to separating funnel and add further diethylether and repeat **(1 mark)** Evaporation/distillation of all diethylether layers **(1 mark)**	3	
		(iii)	Immiscible in water **(1 mark)** Benzocaine soluble in it **(1 mark)**	2	

Question			Expected response	Max mark	Additional guidance
	(b)		recrystallisation **or** chromatography **or** glc	1	
	(c)	(i)	(Pure) benzocaine	1	
		(ii)	Benzocaine has a small impurity. **or** Any other appropriate answer which suggests the benzocaine is not 100% pure.	1	
9.	(a)	(i)	2	1	
		(ii)	1	1	
	(b)		Rate = $k[NO]^2[H_2]$	1	
	(c)		Using experiment 1, $1{\cdot}20 \times 10^{-5} = k(4 \times 10^{-3})^2(1 \times 10^{-3})$ $k = \dfrac{1{\cdot}20 \times 10^{-5}}{(16 \times 10^{-6})(1 \times 10^{-3})}$ $k = 750$ **(1 mark)** $mol^{-2}l^2s^{-1}$ **(1 mark)**	2	
10.	(a)			1	
	(b)		Infrared spectroscopy	1	
	(c)		 and **(1 mark)**	2	Correct full structural, shortened structural or skeletal formulae can all be accepted.
11.	(a)	(i)	Cis refers to the geometric isomer where both substituents are on the same "side" of Pt.	1	
		(ii)	A monodentate ligand forms one dative (covalent) bond (to a central metal atom or ion).	1	

Question			Expected response	Max mark	Additional guidance
	(b)		Moles oxoplatin = $\dfrac{5\cdot00}{334\cdot1}$ = $1\cdot49656 \times 10^{-2}$ Moles asplatin = $1\cdot49656 \times 10^{-2}$ Mass asplatin = $1\cdot49656 \times 10^{-2} \times 484\cdot1$ $\qquad\qquad = 7\cdot245\,g$ **(1 mark)** % yield = $\dfrac{6\cdot36 \times 100}{7\cdot245}$ = $87\cdot8\%$ **(1 mark)**	2	Significant figure rule applies. 88, 87·78, 87·785.
12.			**The whole candidate response should first be read to establish its overall quality in terms of accuracy and relevance to the problem/situation presented.** There may be strengths and weaknesses in the candidate response: *assessors should focus as far as possible on the strengths, taking account of weaknesses (errors or omissions) only where they detract from the overall answer in a significant way, which should then be taken into account when determining whether the response demonstrates* **reasonable**, **limited** *or* **no** *understanding.* **Assessors should use their professional judgement to apply the guidance below to the wide range of possible candidate responses.**	3	This open-ended question requires candidates to outline the possible steps in the synthesis of propanoic acid from other small molecules such as ethene or ethanol. Candidate responses are expected to make comment on the basis of relevant chemistry ideas/concepts which might include one or more of: addition reaction, nucleophilic substitution reaction, reduction reaction, oxidation reaction and hydrolysis reaction.
			3 marks: The candidate has demonstrated a **good** conceptual understanding of the chemistry involved, providing a logically correct response to the problem/situation presented. This type of response might include a statement of principle(s) involved, a relationship or equation, and the application of these to respond to the problem/situation. This does not mean the answer has to be what might be termed an "excellent" answer or a "complete" one.		In response to this question, a **good** understanding might be demonstrated by a candidate response that: • makes comments based on one relevant chemistry idea/ concept, in a **detailed/ developed** response that is **correct or largely correct** (any weaknesses are minor and do not detract from the overall response) **or** • makes comments based on a range of relevant chemistry ideas/concepts, in a response that is **correct or largely correct** (any weaknesses are minor and do not detract from the overall response) **or** • otherwise demonstrates a good understanding of the chemistry involved.

Question			Expected response	Max mark	Additional guidance
			2 marks: The candidate has demonstrated a **reasonable** understanding of the chemistry involved, showing that the problem/situation is understood. This type of response might make some statement(s) that is/are relevant to the problem/situation, for example, a statement of relevant principle(s) or identification of a relevant relationship or equation.		In response to this question, a **reasonable** understanding might be demonstrated by a candidate response that: • makes comments based on one or more relevant chemistry idea(s)/concept(s), in a response that is **largely correct** but has **weaknesses** which detract to a small extent from the overall response **or** • otherwise demonstrates a reasonable understanding of the chemistry involved.
			1 mark: The candidate has demonstrated a **limited** understanding of the chemistry involved, showing that a little of the chemistry that is relevant to the problem/situation is understood. The candidate has made some statement(s) that is/are relevant to the problem/situation.		In response to this question, a **limited** understanding might be demonstrated by a candidate response that: • makes comments based on one or more relevant chemistry idea(s)/concept(s), in a response that has **weaknesses** which detract to a large extent from the overall response **or** • otherwise demonstrates a limited understanding of the chemistry involved.
			0 marks: The candidate has demonstrated **no** understanding of the chemistry that is relevant to the problem/situation. The candidate has made no statement(s) that is/are relevant to the problem/situation.		Where the candidate has only demonstrated knowledge and understanding of chemistry **that is not relevant to the problem/situation presented**, 0 marks should be awarded.
13.	**(a)**	**(i)**	Empirical formula $\quad$ C $\quad$ H $\quad$ O $\dfrac{79{\cdot}25}{12}\quad\dfrac{5{\cdot}66}{1}\quad\dfrac{15{\cdot}09}{16}$ $\dfrac{6{\cdot}60}{0{\cdot}94}\quad\dfrac{5{\cdot}66}{0{\cdot}94}\quad\dfrac{0{\cdot}94}{0{\cdot}94}$ $\quad$ **(1 mark)** $\quad$ 7 : 6 : 1 C_7H_6O $\qquad\qquad$ **(1 mark)**	2	1 mark for calculating number of moles. 1 mark for formula.
		(ii)	C_7H_6O	1	

Question			Expected response	Max mark	Additional guidance
	(b)		$C=O$ or carbonyl group	1	
	(c)			1	Correct full structural, shortened structural or skeletal formulae can all be accepted. Name is not required.

ADVANCED HIGHER FOR CfE CHEMISTRY
MODEL PAPER 1

SECTION 1

Question	Response	Mark
1.	D	1
2.	D	1
3.	C	1
4.	C	1
5.	A	1
6.	A	1
7.	C	1
8.	B	1
9.	C	1
10.	A	1
11.	B	1
12.	A	1
13.	B	1
14.	A	1
15.	B	1
16.	B	1
17.	A	1
18.	D	1
19.	C	1
20.	B	1
21.	A	1
22.	B	1
23.	C	1
24.	D	1
25.	B	1
26.	C	1
27.	B	1
28.	D	1
29.	D	1
30.	D	1

SECTION 2

Question			Expected response	Max mark	Additional guidance
1.	(a)		Energy is absorbed resulting in electrons being promoted to a higher energy level.　**(1 mark)** The line is produced from the electron falling to its original level, emitting energy which corresponds to the wavelength (or frequency) of one of the lines.　**(1 mark)**	2	
	(b)		$E = \dfrac{Lhc}{(1000)\,\lambda}$ or $= \dfrac{6 \cdot 02 \times 10^{23} \times 6 \cdot 63 \times 10^{-34} \times 3 \times 10^{8}}{585 \times 10^{-9} \times (1000)}$　**(1 mark)** $= 205\,\text{kJ mol}^{-1}$　**(1 mark)**	2	Alternatively, calculate the frequency first using $c = f\lambda$ and then calculate energy using $E = Lhf$. i.e. $f = \dfrac{3 \times 10^{8}}{585 \times 10^{-9}}$ $f = 5 \cdot 13 \times 10^{14}\,\text{s}^{-1}$ $E = Lhf$ $\quad = 6 \cdot 02 \times 10^{23} \times 6 \cdot 63$ $\quad \times 10^{-34} \times 5 \cdot 13 \times 10^{14}$ $\quad = 204\,680\,\text{J mol}^{-1}$
	(c)		18 ppm = 18 mg per kg 　10 kg = 180 mg of Ne = 180×10^{-3} g　**(1 mark)** 　Moles $= \dfrac{180 \times 10^{-3}\,\text{g}}{20 \cdot 2} = 0 \cdot 00891$ mol　**(1 mark)** 　Volume = moles × molar volume 　　　　$= 0 \cdot 00891 \times 22 \cdot 4$ litres 　　　　$= 0 \cdot 2$ litres　**(1 mark)**	3	Even if you forget the relationship between ppm and mass, you would still gain credit for attempting to calculate a mass (and hence, number of moles) and showing that you can convert moles into a volume.
2.	(a)		$H_2O(\ell) \rightleftharpoons H^+(aq) + OH^-(aq)$ Methanoate ions react with H^+ ions from water's equilibrium.　**(1 mark)** This causes the water equilibrium to shift to the right-hand side producing excess OH^- and hence pH > 7.　**(1 mark)** **or** Produces an excess of OH^-.	2	
	(b)		$pH = pK_a - \log \dfrac{[\text{acid}]}{[\text{salt}]}$ $pH = pK_a - \log \dfrac{[0 \cdot 25]}{2 \cdot 5}$　**(1 mark)** 　　$= 3 \cdot 75 - (-1)$ 　　$= 4 \cdot 75$　**(1 mark)**	2	

Question			Expected response	Max mark	Additional guidance
3.			This is an open-ended question. Please refer to the specimen paper marking instructions.	3	**Suggestions from the author** You could show your knowledge of chemistry by the following: • discussing what is meant by ionisation energy and illustrate with equations • explaining how increasing the nuclear charge pulls electrons closer to the nucleus • using examples of elements in a period which follow the pattern suggested by the student • discussing the exceptions to this rule and explaining in terms of their electronic configurations. For example: Be $1s^2 2s^2$ B $1s^2 2s^2 2p^1$ B has a lower 1st IE than Be. Boron's outer electron is in a 2p orbital rather than a 2s. 2p is further from the nucleus than 2s. The increased distance results in a reduced attraction and so a reduced ionisation energy. In addition, the 2p orbital is screened by the $1s^2$ electrons AND by the $2s^2$ electrons. This reduces the pull from the nucleus and so lowers the ionisation energy.
4.	(a)		No. of moles of thiosulfate = 0.01525×0.102 = 1.56×10^{-3} so moles Cu^{2+} = 1.56×10^{-3} **(1 mark)** Mass Cu per sample = $63.5 \times 1.56 \times 10^{-3}$ $\qquad = 9.88 \times 10^{-2}$ **(1 mark)** Mass of Cu in key = $9.88 \times 10^{-2} \times \dfrac{1000}{25}$ $\qquad = 3.95\,g$ **(1 mark)**	3	Remember, at AH you must not round at intermediate steps, e.g. 3.87 g or 3.88 g. (2 out of 3 marks would be obtained from using 0.1 instead of 0.102.)

Question			Expected response	Max mark	Additional guidance
	(b)		Any of the following: • use distilled/deionised water • rinsings • carry out replicates/duplicates • cover beaker with watch glass when key is being dissolved • increase sample size for titration.	1	The following answers would receive zero: • more titrations/more samples • measure the 10 g of KI accurately • use a more accurate balance.
	(c)	(i)	Hexaaquacopper(II)	1	Spelling must be correct.
		(ii)	$\left[\begin{array}{c} OH_2 \\ H_2O\,{\cdots}\,\underset{\displaystyle OH_2}{\overset{\displaystyle |}{Cu}}\,{\cdots}\,OH_2 \\ H_2O \qquad OH_2 \end{array}\right]^{2+}$	1	You must show that the copper bonds to the O of water. The square brackets, 2+ charge and octahedral shape must also be shown.
		(iii)	An energy gap is created between the d orbitals **or** degeneracy is broken (when ligands bond). **(1 mark)** This energy gap corresponds to a frequency of light which is absorbed. **(1 mark)** Changing the ligand changes the energy gap between the d orbitals resulting in a different frequency of light being absorbed. **(1 mark)**	3	
5.	(a)		$\Delta S = \Sigma S°$ products $- \Sigma S°$ reactants $= (213 \cdot 8 + 72 \cdot 1) - (112)$ **(1 mark)** $= 173 \cdot 9\ J\,K^{-1}\,mol^{-1}$ **(1 mark)**	2	
	(b)		$\Delta G° = \Delta H° - T\,\Delta S°$ Just feasible when $\Delta G° = 0$ **(1 mark)** $\Delta H° - T\,\Delta S° = 0$ $\quad T\,\Delta S° = \Delta H°$ $\quad\quad T = \dfrac{266\,000}{173 \cdot 9}$ **(1 mark)** $\quad\quad\ = 1529 \cdot 6\ K$ **(1 mark)**	3	Since $\Delta H°$ is in kJ and $\Delta S°$ is in J, you must change either the enthalpy or entropy so that you are using the same energy units. In the marking scheme, the enthalpy has been ×1000 so that it has units in J.
6.	(a)		+5 or 5	1	3×-2 for O = −6. Since the overall charge is −1, Br must be +5.
	(b)	(i)	1st order for Br^- **(1 mark)** 2nd order for H^+ **(1 mark)**	2	
		(ii)	Rate = $k[BrO_3^-][Br^-][H^+]^2$	1	k must be lowercase since capital K represents the units of temperature (Kelvin).

Question			Expected response	Max mark	Additional guidance
		(iii)	8 (1 mark) Units: $l^3\,mol^{-3}\,s^{-1}$ (1 mark)	2	The answer for k is obtained by rearranging the equation in (ii) and substituting the data from any of the experiments.
7.	(a)		Hexa-2,4-diene	1	
	(b)		Any correct structure for deca-2,4,6,8-tetraene such as 	1	
	(c)		Conjugated/conjugation	1	
	(d)		As the number of alternating single and double bonds increases, the energy difference between the HOMO and the LUMO decreases.	1	
	(e)		α-carotene has a shorter conjugated system or similar.	1	
	(f)		 Either of the two carbons connected to an OH group **or** section with one of the asymmetric carbon atoms circled.	1	Look for a C atom that is connected to four different atoms/groups of atoms.
8.	(a)		Ether	1	
	(b)		Add an alkali metal to methanol	1	
	(c)	(i)	First step – show the heterolytic fission of the C—Cl bond to form the carbocation. Second step – show the nucleophilic attack of the methoxide ion. (**1 mark** for correct reactant and for curly arrow showing heterolytic fission of C—Cl bond **1 mark** for correct carbocation intermediate **1 mark** for second curly arrow and final product)	3	

Question			Expected response	Max mark	Additional guidance
		(ii)	Should mention one of two concepts: **either** the carbocation formed has three methyl (alkyl) groups attached which can feed in electron density stabilising the positive charge thus making an S_N1 mechanism more favourable **or** the tertiary haloalkane has three methyl groups attached which offer steric hindrance w.r.t. the formation of the five co-ordinate transition state seen as part of the S_N2 mechanism; bulky groups/ too crowded/steric hindrance.	1	
	(d)	(i)	An equal mixture of enantiomers/both optical isomers	1	
		(ii)	Nothing	1	One enantiomer would rotate the polarised light in one direction; the other enantiomer would rotate the light in the opposite direction. As there is an equal amount of each enantiomer, the rotations would cancel.
9.	(a)		Fairly high molecular mass/available in high purity/thermodynamically stable/soluble in water	1	
	(b)		1. Calculate the mass needed. 2. Accurately weigh required mass of KIO_3 and dissolve completely in small volume of water. 3. Transfer the solution to a standard flask, rinsing the beaker with deionised water and transferring the rinsings to the flask. 4. Add deionised water up to the mark adding the last few drops with a dropper. 5. Invert to mix.	2	
	(c)		Moles of KIO_3 required = CV = $0.1 \times 0.1 = 0.01$ Volume of $1.00\,mol\,l^{-1}$ solution required $= \dfrac{moles}{C}$ $= \dfrac{0.01}{1} = 0.01$ litres Measure $10\,cm^3$ of $1.00\,mol\,l^{-1}$ stock solution using a pipette and transfer to a $100\,cm^3$ standard flask. Add deionised water up to the mark, stopper and invert. **(1 mark for correct volume; 1 mark for correct procedure)**	2	

Question			Expected response	Max mark	Additional guidance
10.	(a)		Condensation	1	
	(b)		CH_3CH_2CN or H—C—C—C≡N (with H atoms)	1	
	(c)		(structure with Hc circled) Hc is the three H atoms on the left-hand side of the molecule.	1	

Question			Expected response	Max mark	Additional guidance
	(d)		This is an open-ended question. Please refer to the specimen paper marking instructions.	3	**Suggestions from the author** IR: Discuss how IR can identify functional groups. Quote some of the wavenumbers from the data booklet for the compounds of interest, i.e. for conversion from an alcohol and carboxylic acid into an ester you would expect to see a decrease/disappearance of —OH from the alcohol and acid and an appearance of the ester functional group. This could be done by sampling at fixed periods of time and observing the change in IR spectrum as the reaction proceeds. Could discuss comparing the reaction mixture with reference samples/library spectra of the product ester. Chromatography: Could run a TLC of three standards, i.e. methanol, propanoic acid and methyl propanoate to calculate/observe their Rf. Run plates with these three spots alongside samples from the reaction to see the appearance of a spot corresponding to the ester (compare with the reference). Discuss the need for spotting agents. Discuss the pros and cons of each technique.

Question			Expected response	Max mark	Additional guidance
11.	(a)		Condensor Heating mantle	2	
	(b)	(i)	To purify (the sulfanilamide) To get rid of impurities	1	
		(ii)	Insoluble in cold water Soluble in hot water	1	
		(iii)	Vaccuum filtration	1	
	(c)		GFM reactant 214·1, GFM product 172·1 Both GFM calculated correctly　　　　(1 mark) $\frac{4\cdot282}{214\cdot1} \times 172\cdot1 = 3\cdot442$　　　　(1 mark) $\frac{2\cdot237}{3\cdot442} \times 100 = \mathbf{65\%}$　　　　(1 mark) or $\frac{4\cdot282}{214.1} = 0\cdot02$ moles $\times 172\cdot1 = 3\cdot442$　(2 marks) Then $\frac{2\cdot237}{3\cdot442} \times 100 = 65\%$　　　　(1 mark)	3	
	(d)		The sample is mixed with pure sulfanilamide　　　　(1 mark) The melting point of the mixture will be the same　　　　(1 mark)	2	
	(e)		Chromatography or IR or NMR or TLC	1	

ADVANCED HIGHER FOR CfE CHEMISTRY
MODEL PAPER 2

SECTION 1

Question	Response	Mark
1.	A	1
2.	A	1
3.	D	1
4.	D	1
5.	B	1
6.	B	1
7.	B	1
8.	C	1
9.	C	1
10.	D	1
11.	D	1
12.	C	1
13.	D	1
14.	C	1
15.	D	1
16.	D	1
17.	B	1
18.	B	1
19.	A	1
20.	A	1
21.	C	1
22.	A	1
23.	C	1
24.	A	1
25.	C	1
26.	D	1
27.	D	1
28.	C	1
29.	C	1
30.	B	1

SECTION 2

Question			Expected response	Max mark	Additional guidance
1.	(a)			1	
	(b)	(i)	$n = 3$, $l = 0$, $m = 0$, $s = +\frac{1}{2}$ or $-\frac{1}{2}$	2	The first quantum number (n) refers to the shell the electron is in. In this case, it is in the 3rd shell so $n = 3$. The second quantum number (l) refers to the orbital, i.e. is it an s, p, d etc. In this case, it is an s orbital which is known as $l = 0$. If it was a p orbital, l would be 1; if it was a d orbital, l would be 2. The third quantum number (m) for an s orbital is always 0. For p orbitals, m can be –1, 0 and +1 (px, py and pz). The final quantum number (s) refers to the spin of the electron. It is either $+\frac{1}{2}$ or $-\frac{1}{2}$.
		(ii)	The 1s orbital has the same shape as the 2s orbital but is smaller **or** has a lower energy than the 2s orbital.	1	
		(iii)	The two electrons occupy degenerate 3p orbitals singly and with parallel spins.	1	
		(iv)	4d	1	
	(c)		5000 ppm = 5000 mg per kg 2 kg = 10 000 mg of CH_4 = 10 g **(1 mark)** Moles = $\frac{10\,g}{16}$ = 0·625 mol **(1 mark)**	2	
2.	(a)	(i)	Moles EDTA = 0·11 × 0·001175 = 1·29 × 10⁻³ Moles Ni = 4 × moles EDTA = 5·16 × 10⁻³ **(1 mark)** Mass Ni = 5·16 × 10⁻³ × 58·7 = 0·303 g **(1 mark)** % Ni = $\frac{0·303}{1·33}$ × 100 = 22·8% **(1 mark)**	3	1 mole of EDTA reacts with 1 mole of the Ni ions. The number of moles calculated has to be ×4 since only 25 cm³ of the original 100 cm³ solution was used.

Question			Expected response	Max mark	Additional guidance
		(ii)		1	To produce a violet colour, red and blue must be mixed, i.e the sample must not absorb red or blue. Assuming the three colours which mix to make white light are red, green and blue, the sample must absorb strongly in the green (509 nm) region but allow all red (~650 nm) and blue (~480 nm) to transmit. The data booklet can be used to determine the approximate λ of red, green and blue.
	(b)		Titration: end point uncertainty **(1 mark)** Colorimetry: reliability of calibration graph or dilute solutions giving low absorbance **(1 mark)** Gravimetric: small mass of sample giving large uncertainty in balance readings/mass transfer or mechanical losses **(1 mark)**	3	These are the sources of most significant error. You may be tempted to mention apparatus error, but this is not a major factor except for gravimetric analysis where the ability of the balance to measure accurate masses is crucial.
	(c)		$E = \dfrac{Lhc}{(1000)\,\lambda}$ **or** $= \dfrac{6\cdot02 \times 10^{23} \times 6\cdot63 \times 10^{-34} \times 3 \times 10^{8}}{508 \times 10^{-9} \times (1000)}$ **(1 mark)** $= 236\,\text{kJ mol}^{-1}$ **(1 mark)**	2	Alternatively, calculate the frequency first using $c = f\lambda$ and then calculate energy using $E = Lhf$. i.e. $f = \dfrac{3 \times 10^{8}}{508 \times 10^{-9}}$ $f = 5\cdot91 \times 10^{14}\,\text{s}^{-1}$ $E = Lhf$ $= 6\cdot02 \times 10^{23}$ $\times\, 6\cdot63 \times 10^{-34}$ $\times\, 5\cdot91 \times 10^{14}$ $= 235\,704\,\text{J mol}^{-1}$

The graph in (ii) shows Absorption (y-axis) against Wavelength/nm (x-axis, marked 400, 500, 600, 700), with a curve peaking around 550–560 nm.

Question			Expected response	Max mark	Additional guidance
3.			This is an open-ended question. Refer to the marking scheme from the specimen paper.	3	**Suggestions from the author** You should use lots of examples to illustrate your understanding of colour. You could discuss: • the colour in transition metal compounds caused by loss of d orbital degeneracy • the conjugation in organic molecules causing electron movement between HOMO and LUMO • the emission of photons from atoms as a result of energy being absorbed which corresponds to the energy gap in an atom. If you can quote specific examples of coloured compounds (and the reasons for the colour), use them!
4.	(a)		$K = \dfrac{[CH_4][H_2S]^2}{[CS_2][H_2]^4}$	1	
	(b)		281	1	
5.	(a)		$HCO_3^-(aq)$ **or** HCO_3^-	1	
	(b)		$pH = \dfrac{1}{2}pK_a - \dfrac{1}{2}\log c$ Substitute values: $pH = \dfrac{1}{2}(6\cdot35) - \dfrac{1}{2}\log 0\cdot 01$ **(1 mark)** $pH = 4\cdot18$ **(1 mark)**	2	
	(c)	(i)	Any indicator with a pH change in the alkaline region. From the data booklet, the following would be acceptable: phenolphthalein, cresol red, thymol blue or thymolphthalein.	1	The solution formed at the end point would contain potassium carbonate which is an alkaline salt.
		(ii)	**1 mark** for stating that a buffer is formed. **1 mark** for explaining that excess H^+ ions are removed by the conjugate base (formed from the potassium carbonate) **or** that excess base reacts with H^+ from the weak acid.	2	A weak acid mixed with the salt of a weak acid forms a buffer solution.
6.	(a)		$-220\cdot3\,J\,K^{-1}\,mol^{-1}$	1	$\Delta S = \Sigma S_{products} - \Sigma S_{reactants}$ $= ((2 \times 27) + 5\cdot7) - ((2 \times 33) + 214)$ $= (59\cdot7) - (280)$ $= -220\cdot3$

Question		Expected response	Max mark	Additional guidance

	(b)	$\Delta H° = -592 \text{ kJ mol}^{-1}$ **(1 mark)**
		
		$\Delta G° = \Delta H° - T\Delta S°$
		
		    $= -592 - 298 \times \left(\dfrac{-220\cdot3}{1000}\right)$ **(1 mark)**
		
		    $= -526\cdot35 \text{ kJ mol}^{-1}$ $(-526 \text{ kJ mol}^{-1})$ **(1 mark)**
			3	To obtain ΔH, use Hess' law to rearrange the equations given,
				i.e. $\Delta H = (2 \times -493) + 394$.
				
				Since this is standard conditions, $T = 298$ K.
				
				Remember to divide the ΔS by 1000 so that all units are in kJ.

7.	**(a)**	3rd order	1	This is obtained by adding the indices from the rate equation, i.e. 2 (from NO) and 1 (from Cl_2).
	(b)	Reaction 3, since the rate equation shows that the rate is independent of the concentration of ammonia.	1	In other words, it doesn't matter what the concentration of ammonia is, the rate is not affected.
	(c)	$k = \dfrac{[\text{Rate}]}{[NO]^2[Cl_2]} = \dfrac{1\cdot43 \times 10^{-6}}{[0\cdot250]^2[0\cdot250]}$ $= 9\cdot15 \times 10^{-5} \text{ l}^2 \text{ mol}^{-2} \text{ s}^{-1}$ **(1 mark for correct value; 1 mark for correct units)**	2	

8.	**(a)**	Moles sulfuric $= 0\cdot1 \times 0\cdot0121 = 1\cdot21 \times 10^{-3}$
		
		Moles of NaOH in 25 cm^3 $= 1\cdot21 \times 10^{-3} \times 2$
		                      $= 2\cdot42 \times 10^{-3}$
		                           (1 mark)
		
		Moles of NaOH in 500 cm^3 $= 20 \times 2\cdot42 \times 10^{-3}$
		                        $= 0\cdot0484$ **(1 mark)**
		
		Moles NaOH added at the start $= 1 \times 0\cdot050$
		                         $= 0\cdot05$
		
		Moles NaOH reacting $= 0\cdot05 - 0\cdot0484$
		                    $= 1\cdot6 \times 10^{-3}$ **(1 mark)**
		
		Moles of aspirin $= \dfrac{1\cdot6 \times 10^{-3}}{2} = 8 \times 10^{-4}$
		
		Mass of aspirin $= 8 \times 10^{-4} \times 180$
		                $= 0\cdot144$ g **(1 mark)**
			4	

	(b)	Aspirin is not very soluble in water.	1	
	(c)	**(i)**	Any of the following for **1 mark**: • mass transfer or mechanical losses • purification of the product results in loses • side reactions • the aspirin hydrolyses to form reactants • impure reactants	1	

Question			Expected response	Max mark	Additional guidance
		(ii)	74% of x = 8 g $x = \dfrac{8}{0\cdot74} = 10\cdot810\,g$ **(1 mark)** Moles of aspirin $= \dfrac{10\cdot810}{180} = 0\cdot060$ Moles of 2-hydroxybenzoic acid $= 0\cdot060$ Mass of 2-hydroxybenzoic acid $= 0\cdot060 \times 138$ $= 8\cdot29\,g$ **(1 mark)**	2	
		(iii)	**1 mark** for melting point. **1 mark** for TLC. **Melting point:** • Compare the melting point obtained with the literature values. A sharp mp close to the literature value would suggest high purity. • A different mp would suggest a different compound. • A lower mp over a long range would suggest the presence of impurities. **TLC:** • A single spot would suggest 1 compound; more than 1 spot would suggest impurities. • Run the TLC with a control (pure sample of aspirin). R_f values for the control should match the sample made.	2	
(d)		(i)	It means that you need to know the exact (accurate) mass of the compound you weigh and it should be in the region of 0.40 g.	1	
		(ii)	100 cm³ flask contains 0·4 g 500 cm³ flask contains 0·4 g Concentration $= \dfrac{mass}{volume} = \dfrac{0\cdot4}{0\cdot5} = 0\cdot8\,gl^{-1}$	1	
		(iii)	Choose a filter where maximum absorbance occurs.	1	
		(iv)	Use a pipette to measure out the 10 cm³ instead of a burette.	1	

Question			Expected response	Max mark	Additional guidance
9.	(a)		Phosphoric acid **or** sulfuric acid **or** aluminium oxide	1	
	(b)		This is an open-ended question. Please refer to the specimen paper marking instructions.	3	**Suggestions from the author** Cyclohexanone is a ketone. You could carry out chromatography, comparing the retention time of the product with that of a pure sample of cyclohexanone. Spectroscopic techniques would give lots of information: Mass spec would give an ion fragment pattern which you could compare to the formula mass of this product. ^{1}H NMR would give a distinctive splitting pattern (you could discuss a likely pattern) or you could compare a spectrum of a pure (or reference) sample of cyclohexanone with this sample, likewise with all spectroscopic techniques. IR-specific carbonyl stretch — quote the data booklet value. Empirical formula — determination from microanalysis.
10.	(a)		Electrophile	1	
	(b)	(i)	 Must be tetrahedral but dots and wedges can be replaced by solid lines.	1	

Question			Expected response	Max mark	Additional guidance
		(ii)	While one group would be able to bind to the appropriate region, the other two would not. **or** The three "functional" groups fail to match the binding regions of the active site. **or** Only one group or two groups could bind (or match) the binding regions. **or** The groups on the lactate ion no longer match the binding regions on the active site of the enzyme. **or** The lactate ion no longer complements the binding region (of the active site). **or** The groups now fail to match the binding region (of the active site).	1	
11.	(a)		HBr **or** hydrogen bromide	1	
	(b)	(i)	Mixing of the s orbital with two p orbitals	1	
		(ii)	Sigma bonds – end-on overlap of (atomic) orbitals Pi bonds – sideways overlap of (atomic) orbitals	1	
	(c)			1	
	(d)		 –ve charge must be given outside the brackets or on the C (in the transition state)	2	

Question			Expected response	Max mark	Additional guidance
12.	(a)	(i)	3·52 g CO_2 contains 0·96 g C 1·44 g H_2O contains 0·16 g H (both for 1 mark) Therefore, 0·64 g O (1 mark)	2	
		(ii)	$C = \dfrac{0·96}{12} = 0·08$; $H = \dfrac{0·16}{1} = 0·16$; $O = \dfrac{0·64}{16} = 0·04$ Ratio = 2 : 4 : 1 Empirical formula is therefore C_2H_4O.	1	
	(b)		$C_4H_8O_2$	1	
	(c)	(i)	 Butanoic acid	2	
		(ii)	Acts as a reference	1	

ADVANCED HIGHER FOR CfE CHEMISTRY
MODEL PAPER 3

SECTION 1

Question	Response	Mark
1.	C	1
2.	C	1
3.	A	1
4.	B	1
5.	C	1
6.	B	1
7.	C	1
8.	B	1
9.	A	1
10.	D	1
11.	C	1
12.	A	1
13.	B	1
14.	C	1
15.	C	1
16.	C	1
17.	C	1
18.	D	1
19.	C	1
20.	B	1
21.	A	1
22.	D	1
23.	C	1
24.	B	1
25.	A	1
26.	A	1
27.	D	1
28.	D	1
29.	A	1
30.	A	1

SECTION 2

Question			Expected response	Max mark	Additional guidance
1.	(a)	(i)	An electron is excited/promoted to a higher energy level. **(1 mark)** When the electron falls to a lower energy level/ground state, energy (corresponding to red light) is emitted. **(1 mark)**	2	
		(ii)	$E = \dfrac{Lhc}{(1000)\,\lambda}$ or $= \dfrac{6 \cdot 02 \times 10^{23} \times 6 \cdot 63 \times 10^{-34} \times 3 \times 10^{8}}{706 \times 10^{-9} \times (1000)}$ **(1 mark)** $= 170 \, kJ \, mol^{-1}$ **(1 mark)**	2	Alternatively, calculate the frequency first using $c = f\lambda$ and then calculate energy using $E = Lhf$ i.e. $f = \dfrac{3 \times 10^{8}}{706 \times 10^{-9}}$ $\quad f = 4 \cdot 25 \times 10^{14} \, s^{-1}$ $\quad E = Lhf$ $\quad\quad = 6 \cdot 02 \times 10^{23} \times 6 \cdot 63 \times 10^{-34}$ $\quad\quad\quad \times 4 \cdot 25 \times 10^{14}$ $\quad\quad = 169\,600 \, J \, mol^{-1}$
	(b)		$1, 0, 0, -\dfrac{1}{2}$	1	In other words, everything is the same except we put a negative sign in front of the $\dfrac{1}{2}$ to show that the spin is in the opposite direction.
	(c)	(i)	 1s 2s 2p 3s 3p	1	Full or half-head arrows would be acceptable.
		(ii)	The (three degenerate) **3p orbitals** are filled singly. or **3p orbitals** are filled in such a way as to maximise the number of parallel spins.	1	
		(iii)	$4, 0, 0, +\dfrac{1}{2}$ or $4, 0, 0, -\dfrac{1}{2}$ or $4, 0, 0, \dfrac{1}{2}$	1	
	(d)		Many **different** electron transitions (in the visible region) or Many **different** energy levels	1	
	(e)		$73 \, mg$ per $kg = 73 \times 10^{-3} \, g$ per $1000 \, g$ $= 7 \cdot 3 \times 10^{-11} \, g$ of Hg in $10^{-6} \, g$ of hair **(1 mark)** Number of mol $= \dfrac{7 \cdot 3 \times 10^{-11}}{200 \cdot 6}$ $\quad\quad\quad\quad\quad\quad = 3 \cdot 64 \times 10^{-13} \, (mol)$ **(1 mark)**	2	

Question			Expected response	Max mark	Additional guidance
2.	(a)		Lone pairs/electrons available to form dative covalent bonds	1	
	(b)	(i)	III **or** 3 **or** 3+ **or** +3	1	
		(ii)	Oxidising agent/to oxidise Cr(III)/electron acceptor	1	The hydrogen peroxide has caused the chromium to change from Cr^{3+} to Cr^{6+}. In other words, the hydrogen peroxide has caused this oxidation.
	(c)		Hexacyanochromate(II)	1	
3.	(a)			1	
	(b)		Number of moles of $CaSO_4 = \dfrac{3 \cdot 89}{136 \cdot 1} = 0 \cdot 0286$ Number of moles of $H_2O = \dfrac{1 \cdot 05}{18} = 0 \cdot 05831$ **(1 mark)** Value of x = 2 **(1 mark)**	2	
	(c)		Moles of oxalic acid $= 16 \cdot 55 \times 10^{-3} \times 0 \cdot 0563$ $= 9 \cdot 32 \times 10^{-4}$ **(1 mark)** Conc. of NaOH $= \dfrac{9 \cdot 32 \times 10^{-4} \times 2}{0 \cdot 020}$ $= 0 \cdot 0932 \, mol \, l^{-1}$ **(1 mark)**	2	
	(d)		Not available in high purity/not stable in solid **or** Solution/solid is deliquescent/hygroscopic/low gfm/absorbs water/absorbs CO_2	1	
4.	(a)		pH remains/stays the same/doesn't change/reasonably constant if small volumes of acid or alkali are added	1	
	(b)		$pH = pKa - \log \dfrac{[acid]}{[salt]}$ Moles of salt $= \dfrac{2 \cdot 24}{112 \cdot 1} = 0 \cdot 0200$ Conc. of salt $= \dfrac{0 \cdot 0200}{0 \cdot 25} = 0 \cdot 08 \, mol \, l^{-1}$ **(1 mark)** $pH = 4 \cdot 87 - \log \dfrac{0 \cdot 2}{0 \cdot 08}$ **(1 mark)** $4 \cdot 87 - 0 \cdot 398 = 4 \cdot 5$ **(1 mark)**	3	The pH equations are in the SQA data booklet. Make sure you know when to use the appropriate equation.

Question			Expected response	Max mark	Additional guidance
5.	(a)		1st order	1	Compare experiments 1 and 3. Doubling [HI] leads to a doubling of the rate, i.e. the reaction is 1st order w.r.t. HI.
	(b)		Rate = $k[H_2O_2][HI]$	1	If you make a mistake in part (a) you will still gain credit for a correct rate equation which corresponds to the orders you calculated in (a).
	(c)		$k = \dfrac{Rate}{[H_2O_2][HI]} = \dfrac{4 \cdot 3 \times 10^{-9}}{[3 \cdot 2 \times 10^{-4}][4 \cdot 1 \times 10^{-4}]}$ $= 0 \cdot 0328 \, mol^{-1} \, l \, s^{-1}$ **(1 mark for answer; 1 mark for units)**	2	The value for k is calculated by rearranging the rate equation and substituting in data from any horizontal line in the table. Here, line 1 is used.
6.	(a)		$178 \cdot 3 \, kJ \, mol^{-1}$	1	Calculated from $\Sigma Hp - \Sigma Hr$ $= (-635 \cdot 1 + -393 \cdot 5) - (-1206 \cdot 9)$
	(b)		$159 \, J \, K^{-1} \, mol^{-1}$	1	Calculated from $\Sigma Sp - \Sigma Sr$ $= (38 \cdot 1 + 213 \cdot 8) - (92 \cdot 9)$
	(c)		$\Delta G° = \Delta H° - T \, \Delta S°$ Just feasible when $\Delta G° = 0$ **(1 mark)** $\Delta H° - T \, \Delta S° = 0$ $T \, \Delta S° = \Delta H°$ $T = \dfrac{178 \cdot 3 \times 1000}{159}$ **(1 mark)** $= 1121 \cdot 4 \, K$ **(1 mark)**	3	Since $\Delta H°$ is in kJ and $\Delta S°$ is in J, you must change either the enthalpy or entropy so that you are using the same energy units. In the marking scheme, the enthalpy has been ×1000 so that it has units in J.
7.	(a)		$LiAlH_4$	1	
	(b)		$C_8H_9O_2N$	1	
	(c)	(i)	Moles of paracetamol $= \dfrac{1 \cdot 10}{151} = 0 \cdot 0073$ **(1 mark)** 60% of x = 0·073 $x = \dfrac{0 \cdot 0073}{0 \cdot 6} = 0 \cdot 0121$ moles **(1 mark)** Moles of 4AP = 0·0121 Mass of 4AP = 0·0121 × 109 = 1·32 g **(1 mark)**	3	The formula for 4-aminophenol is C_6H_7ON which has a gfm of 109.
		(ii)	The impure paracetamol should be dissolved in (the minimum volume of) hot solvent. **(1 mark)** This hot solution should be filtered and the filtrate cooled. **(1 mark)**	2	

Question		Expected response	Max mark	Additional guidance
	(d)	condenser paracetamol sulfuric acid/water heat **(1 mark for correct apparatus; 1 mark for correct labels)**	2	
8.	(a)	Creates a coloured compound **or** Creates a compound which absorbs in the visible region	1	
	(b)	Solutions of accurate iron concentration are made up and the absorbance of each is measured. **(1 mark)** A calibration curve of concentration vs absorbance is drawn. **(1 mark)** The absorbance of the iron tablet solution is measured and the calibration graph is then used to determine the concentration corresponding to this absorbance. **(1 mark)**	3	
	(c)	Moles of iron = $4{\cdot}67 \times 10^{-3} \times 0{\cdot}1$ $\qquad\qquad\quad = 4{\cdot}67 \times 10^{-4}$ Mass of iron = $4{\cdot}67 \times 10^{-4} \times 55{\cdot}8 = 0{\cdot}0261\,g$ % RDA $= \dfrac{0{\cdot}0261}{0{\cdot}0148} \times 100 = 176\%$	2	1 mark for concept of $55{\cdot}8 \times$ calculated moles of iron. 1 mark for arithmetic.
	(d)	Moles $KMnO_4 = 0{\cdot}002 \times 0{\cdot}0203 = 4{\cdot}06 \times 10^{-5}$ Moles of iron = $2{\cdot}03 \times 10^{-4}$ in $10\,cm^3$ Moles of iron = $2{\cdot}03 \times 10^{-3}$ in $100\,cm^3$ Mass of iron = $2{\cdot}03 \times 10^{-3} \times 55{\cdot}8 = 0{\cdot}113\,g$	3	1 mark for stoichiometry. 1 mark for scaling 1:10. 1 mark for arithmetic.

Question			Expected response	Max mark	Additional guidance
9.	(b)		This is an open-ended question. Please refer to the specimen paper marking instructions.	3	**Suggestions from the author** The functional groups of this molecule suggest that it would be polar. Thus, you could carry out a solvent extraction using water/an alcohol along with a non-polar solvent such as hexane. Shaking these in a separating funnel would allow the toxin to go into the water/alcohol layer. Non-polar compounds would stay in the hexane layer. Perhaps the toxin could be reacted with an alkali (e.g. NaOH) to form a salt which would precipitate from the solution and could be filtered. This could be turned back into the toxin by reaction with an excess of acid. Identification: If solid, mp analysis and compare with data booklet values. Carry out full spectroscopic analysis. IR would show specific peaks – discuss these using the data booklet. Likewise, there would be a unique H NMR pattern – discuss/compare to reference compounds.
10.	(a)		 (1 mark for correct structural formulae for both reactant and product; 1 mark for intermediate)	2	Accept valid equivalent using shortened structural formulae.
	(b)		1-bromobutane would form a primary carbocation which is too unstable.	1	
11.	(a)		Antagonist (1 mark) Blocking the body's natural response (1 mark)	2	
	(b)	(i)	Primary	1	

Question			Expected response	Max mark	Additional guidance
		(ii)	Amines are weak bases	1	
		(iii)		1	
12.	(a)		Carbonyl	1	
	(b)		$C_2H_5^+$ or CHO^+	1	
	(c)		$3:2:1$	1	
	(d)			1	Mass spec shows gfm of 58. Fits the NMR ratio of $3:2:1$ and the peak for the aldehyde H on the NMR is consistent with the expected chemical shift as per the data booklet.
13.			This is an open-ended question. Please refer to the specimen paper marking instructions.	3	**Suggestions from the author** This is an opportunity for you to display your knowledge of hybridisation. You could start your answer by giving the electronic configuration for C: $1s^2 2s^2 2p^2$. Orbital box notation: This would limit C atoms to two bonds. Hybridisation involves: • promotion of a 2s electron to the vacant p orbital • mixing the 2s and 2p orbitals to create hybrid orbitals of equal energy. You could then discuss the different types of hybrids using examples to illustrate your answer. For example: • Alkanes: mixing of the s with three p orbitals to create sp^3 hybrids. Discuss how these overlap in a molecule such as ethane: overlap of sp^3 with sp^3 to create the C–C sigma bond; overlap of sp^3 with s orbital from H to form the C–H sigma bonds.

Question	Expected response	Max mark	Additional guidance
			  • Alkenes: discuss an alkene such as ethene and explain how the unhybridised p orbitals overlap to form pi bonds.